INDIGENOUS MANAGEMENT OF WETLANDS

KING'S SOAS STUDIES IN DEVELOPMENT GEOGRAPHY

Series Editors:
Robert W. Bradnock and Kathy Baker-Smith

Both the School of Oriental and African Studies and King's College, whose geography departments have recently merged, have established international reputations for their research into these areas. This series publishes original research into all aspects of geography in the developing world, particularly linking environmental and development issues. It will be of critical interest to geographers and academics in the fields of development studies, political science, environmental studies and economics.

Also in the series

Environment, Knowledge and Gender
Local Development in India's Jharkhand
Sarah Jewitt

NGO Field Workers in Bangladesh
Mokbul Morshed Ahmad

Perspectives of the Silent Majority: Air Pollution, Livelihood and Food Security
Indepth Studies through PRA Methods on Community Perspectives in Urban and Peri-Urban Areas of Varanasi and Faridabad, India
Amitava Mukherjee

Wide Crossing
The West Africa Rice Development Association in Transition, 1985-2000
John R. Walsh

Water Stress: Some Symptoms and Causes
A Case Study of Ta'iz, Yemen
Chris D. Handley

Global Thinking and Local Action
Agriculture, Tropical Forest Loss and Conservation in Southeast Nigeria
Uwem E. Ite

A Clash of Paradigms: Intervention, Response and Development in the South Pacific
Susan Maiava

Indigenous Management of Wetlands
Experiences in Ethiopia

ALAN B. DIXON
The University of Huddersfield

LONDON AND NEW YORK

First published 2003 by Ashgate Publishing

Reissued 2018 by Routledge
2 Park Square, Milton Park, Abingdon, Oxon OX14 4RN
711 Third Avenue, New York, NY 10017, USA

Routledge is an imprint of the Taylor & Francis Group, an informa business

Copyright © Alan B. Dixon 2003

Alan B. Dixon has asserted his right under the Copyright, Designs and Patents Act, 1988, to be identified as author of this work.

All rights reserved. No part of this book may be reprinted or reproduced or utilised in any form or by any electronic, mechanical, or other means, now known or hereafter invented, including photocopying and recording, or in any information storage or retrieval system, without permission in writing from the publishers.

Notice:
Product or corporate names may be trademarks or registered trademarks, and are used only for identification and explanation without intent to infringe.

Publisher's Note
The publisher has gone to great lengths to ensure the quality of this reprint but points out that some imperfections in the original copies may be apparent.

Disclaimer
The publisher has made every effort to trace copyright holders and welcomes correspondence from those they have been unable to contact.

A Library of Congress record exists under LC control number: 2002111307

ISBN 13: 978-1-138-74295-6 (hbk)
ISBN 13: 978-1-138-74291-8 (pbk)
ISBN 13: 978-1-315-18201-8 (ebk)

Contents

List of Figures		*vii*
List of Tables		*xi*
Preface		*xiii*
Abbreviations		*xv*
1	Introduction	1
2	The Management of Wetland Resources	9
3	Indigenous Knowledge and Wetland Management	31
4	Wetland Resources in Illubabor	51
5	The Research Approach	75
6	The Study Wetlands	95
7	The Hydrology of Valley Bottom Wetlands	121
8	Indigenous Wetland Management in Illubabor	149
9	Indigenous and Scientific Wetland Knowledge	181
10	Sustainable Hydrological Management of Wetlands	205
Bibliography		*225*
Index		*241*

List of Figures

Figure 4.1	The location of Ethiopia and the Ethiopian highlands	52
Figure 4.2	The seasonal pattern of rainfall in Ethiopia	53
Figure 4.3	The location of Illubabor zone within Ethiopia	55
Figure 4.4	Administrative divisions (*weredas*) within Illubabor zone	55
Figure 4.5	The seasonal pattern of rainfall and temperature recorded at Metu, central Illubabor (1967 – 1997)	58
Figure 4.6	The dominant vegetation in Illubabor zone	58
Figure 4.7	Estimates of the total area of wetland under cultivation (ha) in Illubabor zone since 1990	69
Figure 4.8	Hurumu wetland under cultivation in 1994	70
Figure 4.9	Hurumu wetland in a degraded state during 1997	70
Figure 4.10a	Traditional wetland management	72
Figure 4.10b	The intensification of wetland utilization	72
Figure 5.1	The location of the study area within Illubabor	76
Figure 5.2	Summary of the classification and study site selection process	78
Figure 5.3	The dipwell apparatus	81
Figure 5.4	Farmers construct a resource map of Bake Chora wetland using a variety of natural materials	91
Figure 5.5	Field sketch of a seasonal calendar (including rainfall) produced by farmers at Dizi wetland	91
Figure 6.1	The final classification of wetlands into three size categories	96
Figure 6.2	The inflow – outflow characteristics of wetlands in the study area	96
Figure 6.3	Wetland order in the study area	97
Figure 6.4	Wetland shape in the study area	97
Figure 6.5	A conceptual model of wetland development based on the observation of wetland characteristics in the study area	99
Figure 6.6	Location of the study wetlands within the study area	101
Figure 6.7	A wetland agricultural calendar showing the timing of the main land use options	105
Figure 6.8	Chebere wetland	107
Figure 6.9	Wangeneye wetland (under maize cultivation)	107
Figure 6.10	Bake Chora wetland	110
Figure 6.11	Hurumu wetland	110
Figure 6.12	Tulube wetland (1996)	113
Figure 6.13	Tulube wetland (1999)	113
Figure 6.14	Dizi wetland	116
Figure 6.15	Anger wetland	116
Figure 6.16	Supe wetland	118

Figure 7.1	The average monthly rainfall in the study area (central Illubabor)	122
Figure 7.2	The average monthly rainfall (Dizi, Sor and Gore) during the study period (August 1997 – July 1998)	122
Figure 7.3	Monthly rainfall during the study period recorded at each gauge	123
Figure 7.4	Rainfall during the study period compared to the 31 year average of Metu, Dizi, Gore and Sor	123
Figure 7.5	The general trend in water table levels during the study period	125
Figure 7.6	Mean weekly water table elevation in the currently undrained study wetlands (August 1997 – July 1998)	127
Figure 7.7	Mean weekly water table elevation in the currently drained and degraded study wetlands (August 1997 – July 1998)	127
Figure 7.8	The range of mean weekly water table elevations recorded at each study wetland (August 1997 – July 1998)	128
Figure 7.9	The spatial distribution of wetland hydrological characteristics in the study wetlands	131
Figure 7.10	The relationship between dipwell groups	133
Figure 7.11	Mean weekly pH levels in the study wetlands	138
Figure 7.12	Range of pH values in each study wetland (August 1997– July 1999)	138
Figure 7.13	Mean weekly pH levels recorded at the top and bottom of each wetland (August 1997 – July 1999)	139
Figure 7.14	The mean weekly electrical conductivity in the study wetlands (August 1997 – July 1999)	140
Figure 7.15	The mean weekly electrical conductivity at the top and bottom of the study wetlands (August 1997 – July 1999)	140
Figure 7.16	Mean monthly nitrate and phosphate levels recorded in the study wetlands (August 1997 – July 1999)	142
Figure 8.1	Farmer perceptions of rainfall at each wetland	151
Figure 8.2	Farmer perceptions of water table elevation at each site (cultivated)	153
Figure 8.3	Farmer perceptions of water table elevation at each site (uncultivated)	154
Figure 8.4	The seasonal calendar of wetland farming activities produced by Supe farmers	161
Figure 8.5	The practice of ditch blocking as a means of regulating water supply to the wetland (pictured here at Wangeneye)	163
Figure 8.6	The head of Bake Chora wetland showing a range of land uses including the reservation of *cheffe*	166
Figure 8.7	*Cheffe* reservation alongside maize cultivation in Anger wetland	167

List of Figures

Figure 8.8	Dizi farmers' wetland management knowledge and its origins	174
Figure 8.9	Indigenous pest management technology? A scarecrow in Bake Chora wetland	177
Figure 9.1	Farmers' perceptions of rainfall compared to rainfall records	183
Figure 9.2	Farmers' perceptions of water table elevation compared to hydrological records	186
Figure 9.3	Farmers' perceptions of water table compared to the hydrological data	189
Figure 9.4	The water table typology in Bake Chora wetland and the location of an area of discoloured maize	193
Figure 9.5	The relationship between farmers' wetland knowledge and that generated by hydrological monitoring	196
Figure 9.6	The mean weekly water table in Bake Chora wetland and the timing of farmers' main hydrological management activities	197
Figure 9.7	The mean weekly water table at Dizi wetland and the timing of farmers' main hydrological management activities	198
Figure 9.8	The mean weekly water table at Wangeneye wetland and the timing of farmers' main hydrological management activities	200
Figure 9.9	Weekly water table levels in lower Wangeneye wetland	200
Figure 10.1	The variable land use within Supe	207
Figure 10.2	A conceptual model of the current situation of cultivation and abandonment in the wetlands	208
Figure 10.3	The degraded wetland of Goma Gabriel wetland near Bure, pictured after several years of complete cultivation in 1996	209
Figure 10.4	Possible strategies for managing wetland regeneration	211
Figure 10.5	A framework for empowering IK resources	217

List of Tables

Table 1.1	The distribution of wetland functions, products and attributes among wetland types	2
Table 2.1	A hydrogeomorphic classification system for wetlands	11
Table 2.2	A wetland typology derived from ecological and geomorphological wetland classifications	13
Table 3.1	The contrasting ideologies of scientific and indigenous knowledge	36
Table 3.2	A framework for incorporating IK in development	43
Table 4.1	Agroclimatic zones of Ethiopia	57
Table 4.2	The percentage of each *wereda* area occupied by wetlands	63
Table 4.3	Guidelines for wetland development as established by the Natural Disaster Prevention Committee (1988)	67
Table 5.1	Checklist of information collected during PRA sessions	92
Table 6.1	The wetland typology results	98
Table 6.2	Study wetlands and their development stages	100
Table 6.3	The location and general characteristics of each study wetland	102
Table 6.4	The hydrological characteristics of each study wetland	103
Table 6.5	The land use characteristics of each study wetland	104
Table 7.1	Summary of the hydrological characteristics of each cluster of dipwells	129
Table 7.2	Number of weeks with surface water at each dipwell	129
Table 7.3	Classification of K_{sat} values in each dipwell according to FAO (1963)	135
Table 7.4	Correlation matrices for mean phosphate and nitrate concentrations	142
Table 7.5	Summary details of chemical concentrations recorded at each wetland	143
Table 7.6	Summary of the impact of agricultural utilization on the wetland hydrological regime	146
Table 8.1	Differences in perceptions of high and low rainfall levels between sites	152
Table 8.2	Differences in farmers' perceptions of high and low water table levels	155
Table 8.3	Tulube farmers' perceptions of water colour	157
Table 8.4	Summary of wetland seasonal farming calendars	162
Table 8.5	Tools utilized in the wetland farming system	164

Preface

The research on which this book is based was initiated in 1996 in response to concerns that wetlands in Illubabor zone, south-west Ethiopia, were increasingly threatened by over-exploitation. At the time, an intensification in drainage and cultivation of these areas as a result of government and NGO policy was seen as the major threat, and the initial aim of the research was to investigate how intensive drainage in particular, affected the hydrological balance and the ability of wetlands to continue to fulfil their roles in ensuring food security. This study, which was essentially hydrological in nature, was to form part of a larger EU funded project, The Ethiopian Wetlands Research Programme (EWRP), which carried out a range of inter-disciplinary investigations into wetlands and their sustainable use in this part of the country.

It soon became clear, however, that the situation was far more complex than initially envisaged. Not all communities were responding to the pressure to use wetlands in this intensive manner, many wetlands also remained uncultivated, some were clearly degraded and some appeared to have been used for over a century with little sign of degradation. Although it maintained a focus on the wetland hydrology, this research consequently evolved and widened its aims to address the ways in which people interacted with these wetlands; examining how wetlands are used by local communities, variation in their hydrological management techniques, the local knowledge on which these are based and how these techniques have developed over time. By understanding these processes, it was deemed possible to identify the potential for the future sustainable use of the wetlands, hence livelihood security for those communities who depend upon their functions and products. This book, in effect, presents the experiences and findings of this research based on 16 months fieldwork undertaken in Illubabor zone between 1996 and 1999.

A great many people participated in and facilitated this research. I am particularly grateful to the farmers of Metu, Yayu-Hurumu and Ale-Didu *weredas* who patiently allowed the insertion of hundreds of plastic tubes into their wetlands, and also the farmers at Bake Chora, Dizi, Supe, Tulube and Anger wetlands, for their contribution to my wetland knowledge system. Hopefully this research has, in some way, empowered and contributed to their wetland knowledge.

The research from inception to completion would not have been possible if it were not for the support and enthusiasm of Professor Adrian Wood at The University of Huddersfield. Dr Anne Jones, Dr Declan Conway, Dr Julia Meaton and Professor Dave Butcher also contributed to the success of the research through their advice and support.

In Ethiopia, a myriad of people provided technical, logistical and social support during the fieldwork. These included Afework Hailu, Dr Patrick Abbot, Tabeje Girmay, Mesfin Ferede and Endale Mamo of the Ethiopian Wetlands Research

Programme. Thanks also go to Debella Dinka, Habtamu Wubshet, Tewoldebirhan Gebre Kidan and Mintasinot Teshome from the Menschen Für Menschen Foundation, and Dr Solomon Mulugeta and Dr Tesfaye Tafesse from Addis Ababa University. In Metu, Mohammed Fili, Tilahun Semu and Getachew Erena, in particular, facilitated the success of the fieldwork.

In the UK I am also grateful for the support of my human ecologist colleagues, particularly Simon, Marie, Jon, Rob and Lance. Above all, my thanks go to Siân, who has maintained a continuous stream of support and encouragement since day one.

Abbreviations

AMR	Average Monthly Rainfall
CIP	Coffee Improvement Project
CSA	Central Statistics Authority
EC	Electrical Conductivity
EIA	Environmental Impact Assessment
EMA	Ethiopian Mapping Authority
EUE	Emergencies Unit for Ethiopia (in UNDP)
EWRP	Ethiopian Wetlands Research Programme
FAO	Food and Agriculture Organization of the United Nations
FDRE	Federal Democratic Republic of Ethiopia
GIS	Geographical Information System
GPS	Global Positioning System
GWT	Groundwater Table
HEIA	High External Input Agriculture
IK	Indigenous Knowledge
IKS	Indigenous Knowledge System
ITCZ	Inter Tropical Convergence Zone
ITK	Indigenous Technical Knowledge
IUCN	World Conservation Union
K_{sat}	Saturated hydraulic conductivity
LEIA	Low External Input Agriculture
MFM	Menschen für Menschen
MoA	Ministry of Agriculture
MoTCD	Ministry of Tea and Coffee Development
MoWME	Ministry of Water, Minerals and Energy
NDPC	Natural Disaster Prevention Committee
NGO	Non Governmental Organization
OECD	Organization for Economic Co-operation and Development
PRA	Participatory Rural Appraisal
RRA	Rapid Rural Appraisal
SCRP	Soil Conservation Research Project
TOT	Transfer of Technology
UN	United Nations
UNDP	United Nations Development Programme

Chapter 1

Introduction

The importance of wetlands

Although often regarded as wastelands in the past, wetlands are today gradually being recognized as one of the world's most valuable ecosystems in terms of the ecological, hydrological and socio-economic roles they fulfil in their natural state (Maltby, 1986; Dugan, 1990; Roggeri, 1998). Their ability to perform and support such a wide range of functions and products has, however, largely been ignored by developers and planners who have regarded the development and exploitation of wetlands as a means of transforming 'wastelands' into economically beneficial and productive areas (Hughes, 1992). Rather than appreciating the fragile inter-dependency of components of natural wetland systems, and the associated benefits linked to each, wetland development has followed a technocratic approach geared towards narrow economic gain. Consequently, the result in many of the world's wetlands has been the exploitation of one wetland function or product at the cost of others.

The ability of wetlands to perform a variety of environmental functions depends on what Gottlich (1977) termed 'ecohydrology', which refers to the complex inter-relationships between the dynamic behaviour of water and the ecological characteristics it produces. The ecohydrological system, which is also influenced by location and geomorphology, determines the individual charactcristics of each wetland and consequently the unique set of functions a wetland is able to provide. The degree to which local populations are dependant upon wetland functions and products also determines each wetland's socio-economic and socio-cultural significance.

Table 1.1 shows the range of functions and products wetlands are potentially able to provide, although these are by no means common to all wetlands. The wide range of ecohydrological conditions which can exist has necessitated the classification of wetlands into different types, ranging from marshes to mangrove swamps. The occurrence of specific types of wetland throughout the world is both spatially and temporally variable, as are the functions and products associated with each. In addition, the importance of each wetland and its functions is also determined by the surrounding physical and socio-economic environment. Nowhere is this significance greater than in the dry parts of the developing world where, according to Adams (1993a), wetlands have a strategic importance for the rural economy out of all proportion to their size, with whole communities being dependant upon their productivity and hydrological benefits.

Table 1.1 The distribution of wetland functions, products and attributes among wetland types (adapted from Dugan, 1990)

	Freshwater Marshes	Estuaries	Floodplains	Lakes	Peatlands
WETLAND FUNCTIONS					
Groundwater recharge	■	O	■	■	●
Groundwater discharge	■	●	●	●	●
Flood control	■	●	■	■	●
Shoreline stabilization	■	●	●	O	O
Sediment retention	■	●	■	■	■
Nutrient retention	■	●	■	●	■
Biomass export	●	●	■	●	O
Storm protection	O	●	O	O	O
Micro-climate stabilization	●	O	●	●	O
Water transport	O	●	●	●	O
Recreation / Tourism	●	●	●	●	●
PRODUCTS					
Forest resources	O	O	●	O	O
Wildlife resources	■	■	■	●	●
Fisheries	■	■	■	■	O
Forage resources	■	■	■	O	O
Agricultural resources	●	O	■	●	●
Water supply	●	O	●	■	●
ATTRIBUTES					
Biological diversity	●	■	■	■	●
Uniqueness to culture	●	●	●	●	●

O - Absent or exceptional
● - Present
■ - Common and important value of that wetland type

Wetlands under threat

The major threat to the existence of wetlands has been their conversion to commercial, residential or industrial sites and their utilization for agriculture following intensive drainage. In most cases these activities have far reaching consequences on a wetland's ability to continue to perform a wide range of functions. The attractiveness of drained wetlands as a farming resource stems from the relatively high fertility of wetland soils, although research has suggested that once the natural depositional processes in a wetland are disrupted, nutrient levels are not sustained and this results in a dramatic fall in soil fertility following initial agricultural utilization (Barrow, 1991). Any benefits of drainage may, therefore, be short-lived and at the cost of permanent ecohydrological changes. These have wider implications for the range of biodiversity a wetland supports and the human communities who may be dependant on specific wetland products and functions.

Economic growth, industrialization and population pressure have precipitated the increasing exploitation of wetland resources as new areas for development are sought (Maltby, 1986). According to Hollis (1990), these driving forces are also aided by public misperceptions of the benefits of wetlands, local scale planning as opposed to catchment management, international subsidies such as the European Common Agricultural Policy, and narrow disciplinary thinking. When viewed in their entirety wetland development initiatives have achieved only modest success, as has often been the case with large-scale dam and irrigation projects in the developing world (Adams, 1992). Furthermore, those groups in society who gain immediate benefits from wetland development are usually different from those who bear the economic losses or environmental impacts associated with such development. Hollis (1990) maintains that only an interdisciplinary approach to the study of wetlands can balance the needs of different parties, taking into account the views of hydrologists, ecologists, engineers, economists and those who gain benefit from the wetlands themselves. In the past, development projects have tended to degrade the true value of wetlands through a lack of interdisciplinary thinking which has resulted in development with only short term gains, that is development which is unsustainable (Maltby, 1986; Dugan, 1990; Turner, 1991; Roggeri, 1998).

Wetlands in the developing world

Whilst there is a general consensus that the world's wetlands are under threat from development, there is some recognition that in the developing world many wetlands have been used in a sustainable manner for generations. The dependence of local communities on wetland functions and products for their subsistence is well documented (Maltby, 1986; Dugan, 1990; Scoones, 1991; Adams, 1992; Woodhouse et al., 2000), yet little research has addressed the issue of indigenous wetland management. In contrast, rural development research has, since the 1970s, become preoccupied with the role of Indigenous Knowledge Systems (IKS) in the development and management of natural resources. Studies have drawn attention to the highly specialized and adapted natural resource management techniques, which have evolved over generations in response to environmental or socio-economic changes (Brokensha et al., 1980; Chambers, 1983; Scoones and Thompson, 1994). In many cases, these practices can be considered sustainable, showing no degradation in the natural resource base. Problems start to occur, however, when the rate of environmental or socio-economic change increases and exceeds the capacity of local communities to adapt their resource management techniques (Farrington and Martin, 1988; World Bank, 1998). The result is environmental degradation.

Rapid change is undoubtedly occurring in many parts of the developing world, especially in terms of population growth and globalization. Among other natural resources, wetlands are under threat as a result of these changes yet an understanding of the ways in which local communities utilize and manage them may be the key to avoiding their destruction. If this indigenous knowledge is recognized by the range of wetland stakeholders there is, perhaps, an opportunity to exchange knowledge and build upon the existing indigenous capacity to adapt to changing circumstances. In this way, wetland management strategies can remain flexible and sustainable so that wetlands can continue to provide a range of benefits to local communities.

Wetland management in south-west Ethiopia

Ethiopia, the focus of the research activities outlined in this book, is one such country to have experienced periods of rapid social, political and environmental change during the last 50 years, which have had a significant impact on the country's natural resources. Drought, soil erosion and deforestation in the Ethiopian highlands have been among the most documented problems, although these have had significant knock-on effects in terms of their environmental impact on other resources throughout the whole country (Alemneh Dejene, 1990; Cross and Millar, 1994; Wood, 1990).

Although less on the front line of environmental degradation than other parts of the country, Illubabor zone in the south-west highlands has, nonetheless, experienced periods of inward migration, resettlement and commercialization, together with an increasingly unpredictable climate in recent years. To date, however, it has managed to retain much of its rich natural resource base, of which wetlands remain a critical component (approximately 1.4 per cent of the total land area).

The main traditional use of wetlands in this part of the country has been the harvesting of sedge for the roofs of local housing and, in some areas, the small-scale drainage and cultivation of wetland margins has also taken place. These wetland margins have usually remained cultivated for a period of four months per year and with the onset of the rains they revert to inundated swampland. It has been suggested that if used in this manner, the small-scale cultivation of wetlands appears is sustainable, with no demonstrable long-term detrimental effects on the environment (Wood, 1995).

Between 1970 and 1990, Illubabor zone experienced increased inward migration and a consequent growth in population pressure, largely as a result of government resettlement programmes (Alemneh Dejene, 1990). At the same time there also appears to have been an increase in the use of valley bottom wetlands. This has involved, in some cases, an extension of wetland cultivation to eight months of the year with the permanent drainage of whole swamps as opposed to only their margins (Tafesse Asres, 1996). Although there are obvious benefits from such forms of valley bottom drainage in terms of agricultural output, some reports from the area have suggested that these are only short-term and that long-term drainage may lead to permanent changes in the hydrological processes and functioning of valley bottom wetlands (Kebede Tato, 1993; Butcher and Wood, 1995). Farmers themselves in Illubabor have recently suggested that after a number of years of complete drainage the wetlands become degraded in that they no longer have the ability to perform their original functions to the same extent and they can only support the rough grazing of cattle.

If the full range of benefits of these wetlands to the local community are to be sustained for future generations there is clearly a need to understand wetland ecohydrological processes and the changes brought about as a result of drainage and more intensive utilization. Perhaps more importantly, there is also an urgent need to understand these wetland changes in the context of community wetland knowledge and the motivating forces, techniques and practices which influence the use of wetlands and which ultimately affect their sustainable management.

This book presents and discusses the findings of a recent research programme carried out by the author, which attempted to address specifically the physical impacts of increased wetland utilization and contribution of community indigenous knowledge to the sustainable management of wetlands in Illubabor zone.

The research aims

The main aim of the research was to determine the extent to which the hydrological management of wetlands for agricultural development can be sustainable in Illubabor zone. This was achieved by focusing on four key objectives:

1. the identification of the general characteristics of valley bottom wetlands found in Illubabor zone;
2. the identification of the impact of agricultural utilization on the hydrological regimes of these valley bottom wetlands;
3. an exploration of the role of indigenous knowledge in wetland resource management and the ways in which wetland knowledge held by local communities is operationalized;
4. the identification of principles for the sustainable hydrological management of Illubabor's valley bottom wetlands.

Sustainable hydrological management, for the purposes of the research, was considered to be characterized by:

1. management of the wetland hydrology so that there is no long-term degradation of the wetland hydrological regime (hydrological sustainability);
2. management of the wetland hydrology so that it supports a range of wetland functions and benefits which meet the needs of the local community and which can be sustained year after year with no degradation in the wetland resource (output sustainability);
3. management of the wetland hydrology which is socially sustainable, i.e. that hydrological management practices and the knowledge on which these are based can adapt and evolve in response to changing pressures through the accumulation, evolution and application of IK, thereby facilitating hydrological sustainability and output sustainability.

Structure of the book

The second chapter sets the scene for the discussion of wetlands in Illubabor zone by drawing upon the existing literature on wetlands in the developing world. It highlights the importance and problems of classifying wetlands, an important prerequisite to any conservation or management policy initiative, and discusses the range of tropical wetland characteristics, their associated functions and available management options, with particular reference to the East African highland environment.

Chapter Three develops the concept of sustainable management and reviews past and current trends in indigenous knowledge (IK) research, and the implications of this for sustainable wetland management. The chapter ends by examining the relationship between IK and wetland management, focusing on the potential of participatory development as wetland management option.

Chapter Four introduces the study area of Illubabor in South-west Ethiopia and the importance of its environmental resources in relation to the rest of Ethiopia. The environmental and socio-economic characteristics of the area are reviewed, providing a background to subsequent discussions on wetlands and their current and historical significance in Illubabor's landscape.

Chapter Five details the methodological background of the research carried out in Illubabor, describing the process of wetland classification, the selection of study wetlands and the choice of field methods used to investigate sustainable dimensions of wetland use.

Chapter Six presents detailed information on the wetlands on which the research was carried out, providing a context and reference for successive chapters.

Chapter Seven presents and analyzes the findings of the pioneering wetland hydrological investigations. It assesses the impact of human interaction with the wetlands on a range of hydrological variables and examines the overall hydrological sustainability of wetland use in the area.

Chapter Eight discusses the findings of the programme of participatory research, which involved the collection of information on farmers' perceptions of wetland hydrological processes, their utilization techniques, practices and wetland management strategies. The chapter draws attention to the similarities and differences in farmers' IK between wetlands, highlighting variations in hydrological perceptions and utilization strategies, along with the recognized constraints and inherent problems of wetland utilization.

In the context of the discussions in Chapter Seven and Chapter Eight, Chapter Nine discusses the relationship between farmers' knowledge of the wetland environment and hydrological reality. The nature of this relationship can be considered the key to attaining the sustainable hydrological management of these wetlands in that it is farmers' knowledge and the extent to which this reflects hydrological reality, which ultimately determines the ways in which wetlands are used. Consequently, this chapter examines the extent to which hydrological reality is perceived, understood and applied by wetland users in their management practices. On the basis of this knowledge and the extent to which it is operationalized, the potential of wetland users to achieve sustainable hydrological management is also discussed.

The final chapter draws together the key findings of the research with respect to the sustainability of wetland use in Illubabor zone. The discussion draws upon the evidence presented in previous chapters to suggest that there are clear principles and practices linked to farmers' knowledge and identified through the hydrological research programme, which can contribute to the sustainable hydrological management of wetlands. Having established the principles of sustainable hydrological management, the chapter goes on to discuss how these principles can be fully operationalized, in particular, through the development and empowerment of indigenous knowledge resources. The chapter concludes with a discussion of the implications of the research findings for sustainable hydrological management throughout Illubabor zone and in the wider context of Ethiopia.

Chapter 2

The Management of Wetland Resources

Introduction

During the past thirty years the image of wetlands has been transformed from one of undervalued, unutilized wastelands whose only utility lies in their conversion to alternative land uses, to the current situation where they are regarded as internationally important natural resources requiring sustainable management, 'wise use' or conservation. Researchers and practitioners are beginning to recognize that wetlands perform a variety of ecological, hydrological and socio-economic functions, which are of benefit to the human population particularly in many parts of the developing world where local communities are dependant on these functions for their very survival. Despite this recognition of their significance, many wetlands in the developing world continue to be under threat from top-down development initiatives which see the exploitation of wetland resources or the conversion of wetlands as a means of coping with demographic and socio-economic pressures.

What are wetlands?

One of the major problems in wetlands research has been defining what constitutes a wetland. Roggeri (1998) suggests that the term 'wetland' has been created relatively recently by researchers in an attempt to group together a wide variety of landscape units whose ecosystems share the fundamental characteristic of being strongly influenced by water. The Ramsar Convention has adopted such an approach and defines wetlands as:

> ...areas of marsh, fen, peatland or water, whether natural or artificial, permanent or temporary, with water that is static or flowing, fresh, brackish or salt, including areas of marine water the depth of which at low tide does not exceed six metres (Davis, 1994, p.3).

Similarly, Maltby (1986) suggests that wetlands are ecosystems whose formation, processes and characteristics are determined by water, emphasizing how hydrological characteristics have a profound effect on the ecology of wetlands. The dynamic and changing relationship between these two variables, Gottlich's (1977) 'ecohydrology', can induce a wide variety of wetland characteristics and the occurrence of specific characteristics together have, for generations, been given a variety of colloquial names, e.g. bog, mire, swamp and fen.

With increasing recognition of the importance of wetlands, there has been a need to identify more accurately and objectively the specific ecohydrological characteristics of the range of wetland landscape units throughout the world. This need to classify wetlands has largely been driven by research itself, which has focused upon the capacity of wetlands to provide critical functions and products. The provision of these functions are inherently linked to a wetland's ecohydrological relationship, hence the classification of wetlands into various types provides an effective tool for planners, developers and managers. By using a classification system, it is argued that the assessment and study of wetlands can become more manageable and less time consuming although the effectiveness of a classification system ultimately depends on the precise ecological or hydrological criteria used in the classification and it is important that these are relevant to the aims of management or research. It is important that a wetland classification strikes a balance between what is considered too general to provide useful information, and what is too specific to facilitate its application and utilization.

Consequently, a range of different wetland classification systems have been developed. Whereas some are based broadly on a combination of ecohydrological characteristics often within specific geographical locations (e.g. that of the IUCN in East Africa, 1996), others deal specifically with one aspect of ecohydrology (e.g. Gilvear and McInnes, 1994; Brinson *et al.*, 1994) depending upon the purposes of classification.

Hydrological classifications

According to Mitsch and Gosselink (1986), hydrology is probably the single most important determinant for the establishment of specific types of wetlands and wetland processes. The major controlling factors in determining the hydrological characteristics of natural wetlands are climate, topography and geology (Lloyd *et al.*, 1993). Climate is influential through the effects of precipitation and evapotranspiration, topography controls the form of wetland development and the geology of an area may control the supply of water and its discharge from a wetland. Each of these three influential factors occur in combination to affect the dynamics of hydrological inundation and subsequently each wetland's general characteristics. Several researchers have proposed wetland classifications systems based primarily on hydrology but also acknowledging the relative influences of climate, topography and geology. Gilvear and McInnes (1994) in particular, propose a hydrological classification based on the wetland water balance equation:

$$S = P + SWI + GWI - ET - SWO - GWO$$

Where:
S = change in storage
P = Precipitation
SWI = Surface water inflow
GWI = Groundwater inflow
ET = Evapotranspiration
SWO = Surface water outflow
GWO = Groundwater outflow

In this classification the elements in the wetland water balance equation can be broadly classed into inflows and outflows and whereas precipitation and evapotranspiration are common to all wetlands, the other variables only apply to certain wetland types. Using different combinations of inflows and outflows, Gilvear and McInnes (1994) were able to describe 12 hydrological classes combining the source of water as one variable (ombrotrophic, minerotrophic, omnitrophic or rheotrophic) and the nature of the water input as the other (seasonal or flushed, the latter referring to the throughflows of surface and groundwater). The utility of this hydrological classification, according to its authors, lies in its use for assessing the hydrological vulnerability of wetlands to water pollution and because it is based on the water balance equation, it can be used to model wetland hydrology. It is also sensitive enough to distinguish between wetlands which may, in other classification systems, be grouped together despite their hydrological differences.

A similar approach has been taken by Brinson *et al.* (1994) who propose a hydrogeomorphic classification where in addition to the sources of water, the movement of water and the geomorphic setting of the wetland are used as classification criteria. The three variables considered in the overall classification are highlighted in Table 2.1. The advantage of this system is that it uses broad, easily identifiable geomorphic settings as the basis for the initial classification, which is then supplemented by the more site specific variables of hydrological and hydrodynamic characteristics. According to the authors, the result is a transferable classification system which facilitates straightforward and accurate wetland descriptions. Furthermore, classification on the basis of these variables can be a significant indicator of the functions each wetland provides. For example, where overland flow is significant in a riverine wetland, the wetland is likely to perform functions associated with the trapping of sediment and nutrient recharge of the wetland soils (e.g. as characterized by floodplains).

Table 2.1 A hydrogeomorphic classification system for wetlands (adapted from Brinson *et al.*, 1994)

Geomorphic setting	Hydrological input	Hydrodynamic characteristics
1. riverine	1. precipitation	1. vertical fluctuations
2. depressional	2. overland flow	2. unidirectional horizontal flow
3. fringe	3. groundwater discharge	3. bidirectional horizontal flow
4. peatlands		

Other examples of hydrologically based classifications include those by Lloyd *et al.* (1993) and Roggeri (1998). The latter, however, identifies geomorphological units as the more significant factor influencing wetland characteristics such as hydrology and nutrient supply. The basis for this classification include the geomorphological units of alluvial lowlands, small valleys, lake shores and depressions.

Another aspect of wetland hydrology which is an important component of the ecohydrological relationship is the classification of wetland hydrochemistry. Apart from a simple distinction between freshwater, saltwater and brackish conditions, wetlands can be classified according to their trophic status, which refers to the quantity of nutrients available in the water (Harper, 1992, Heathwaite, 1994). A nutrient poor aquatic environment is referred to as oligotrophic, whereas nutrient rich water is termed eutrophic. In addition, where normal conditions exist, water is said to be mesotrophic and under conditions of exceptional nutrient loading, water is hypertrophic.

Such a classification is relatively simple although the actual determination of trophic status requires the measurement of indicator parameters such as phosphate or chlorophyll concentrations which relate to the abundance of plant material. Its utility is reserved primarily for those studies dealing with the influence of pollutants on the wetland environment, in particular the effects of nitrate and phosphate contamination (Rast and Holland, 1988), although it remains a significant indicator of the overall wetland ecohydrology, including wetland flora and fauna. For example, with reference to hydrological classification developed by Gilvear and McInnes (1994), the extent to which a wetland is flushed will be linked to the trophic status of the wetland in terms of either replenishing nutrient supply or diluting the effects of pollutants. Where the latter is concerned, the hydrochemical classification of wetlands is particularly useful in assessing the nature and extent of anthropogenic disturbance.

Ecohydrological classifications

Wetland ecological characteristics which are influenced by, and themselves influence, the hydrological characteristics of a wetland, are also widely used as a means of classification. Roggeri (1998) identifies several specific ecological units which can be used to classify wetlands but these are combined with geomorphological units to present a more general wetland typology as shown in Table 2.2. This incorporation of both hydrological and ecological characteristics has been the most commonly used means of classifying wetlands and agencies such as the IUCN and Ramsar have developed systems which rely on the identification of land units characterized by specific ecohydrological conditions.

The fourth meeting of the contracting parties to the Ramsar convention in 1990 identified 39 sub-categories of wetland within the broad categories of Marine and Coastal Wetlands, Inland Wetlands and Man-made Wetlands (Matthews, 1993). Within this classification, wetland types are clearly based on their ecological and hydrological characteristics, although according to Dugan (1990), this range of wetland types can be classified again according to the landscape units (relating to geomorphology) in which each occur:

1. Estuaries, which constitute bodies of water where a river mouth widens into a marine ecosystem, where the salinity is intermediate between salt and fresh

water, and where tidal action is an important bio-physical regulator;
2. Open coasts, which are not subject to the influence of river water and lagoon ecosystems;
3. Floodplains, which include areas of land between a river channel and the valley sides, which are subject to periodic flooding;
4. Freshwater marshes, which occur where groundwater, springs, streams or runoff cause seasonal inundation. Where standing water is permanent, these areas are also classified as swamps;
5. Lakes and Ponds, which occur as a result of several geomorphological processes including folding, faulting, volcanic disturbances, glaciation or fluvial processes;
6. Peatlands, which are characterized by the accumulation of partially decomposed organic matter, which occurs as a result of conditions of low temperature, high acidity, low nutrient supply, waterlogging and oxygen deficiency;
7. Swamp forests, which develop in still waters around lake margins and floodplains.

Table 2.2 A wetland typology derived from ecological and geomorphological wetland classifications (adapted from Roggeri, 1998)

Ecological Units	Geomorphological Units						
	Alluvial lowlands	Headwater lowlands	Small overflow valleys	Lake shores: drawdown areas	Lake shores: shallows	Depressions of river / lake systems	Isolated depressions
Periodically flooded ecosystems							
- Flooded forests	●	O	●	●	O	●	O
- Flooded grasslands	●	●	●	●	O	●	O
- Seasonal shallow water bodies	●	●	●	●	O	●	O
Marshes and Swamps							
- Flooded grasslands	●	●	●	●	●	●	O
- Marshes / herbaceous swamps	●	●	O	●	●	●	O
- Swamp forests	●	O	O	●	O	●	●
- Peat swamps	O	●	O	●	O	●	●
Permanent shallow water bodies	●	O	O	O	O	●	●

●: common   O: rare

This broad classification, however, is restricted to general considerations of the range of wetland landforms rather than individual characteristics, functions and processes. In contrast, the Ramsar classification is specific enough to allow wetlands with similar ecological and hydrological characteristics to be grouped together. In this way it fulfils the aims of the Ramsar Convention Bureau in that it provides a framework for international co-operation for the conservation and 'wise use' of the world's wetlands resources (Davis, 1994). Other international agencies involved with wetlands have also followed a similar approach to wetland classification, in particular the IUCN who have developed ecohydrological typologies but with reference to specific locations such as East Africa (IUCN, 1996).

Spatial and temporal aspects of wetland distribution

Dugan (1990) suggests that wetlands are heterogeneous ecosystems with a diverse range of ecohydrological characteristics. Within one wetland system the range of ecohydrological characteristics may be such that they are indicative of several types of wetland. Furthermore, many wetlands are also subject to temporal variations in climatic and hydrological conditions, rendering classification problematic.

While no single classification system can represent the diverse range of conditions which exist, typologies such as that adopted by Ramsar are useful in that they are based on the occurrence of ecosystems characterized by a specific ecology-hydrology relationship. This relationship is often influenced further by geomorphological factors and in a wider context by geographical location. As a result, the occurrence of specific types of wetlands tends to be limited by the interaction of climate, ecology and geomorphology to a specific geographic area. Knowledge of these influential factors is, therefore, useful in being able to predict the occurrence of wetland conditions and the range of variation. For example, the cool, wet climate of the temperate and sub-arctic zones favour the development of extensive areas of peatland which, according to Mitsch *et al.* (1994), probably account for over half of the world's wetlands. In tropical areas, however, peatlands are relatively scarce and most are located in highland areas which receive abundant rainfall (Hughes, 1996). Similarly, mangrove forests are the tropical and sub-tropical equivalent of temperate saltwater marshes (Hughes, 1992). In view of this uneven distribution of wetland types throughout the world it is useful to narrow the present discussion of wetlands to those types which relate to the climatic zone in which Ethiopia is located, i.e. tropical wetlands.

The importance of wetlands in the tropics

Tropical wetlands are poorly understood yet they represent some of the most important areas of the world's remaining wetlands. Of the 6.4 per cent of the earth's surface occupied by wetlands, 56 per cent (4.8 million km^2) lie in the tropical and sub-tropical region (OECD, 1996). Roggeri (1998) suggests that the abundance of wetlands in tropical regions can be explained by the fact that they have largely escaped the impact of development in contrast to those located elsewhere throughout the world.

Developing countries, which often have insufficient financial resources, low population densities or a different concept of development, predominate tropical regions. Hence over-exploitation and development has mostly been avoided. Whilst this may be true in terms of large scale, western style development initiatives, it is inaccurate to suggest that wetlands in the developing world have avoided any form of development. Many of these wetlands have undergone various forms of small-scale indigenous development which have been sustained by local communities for generations with no serious environmental impact (Richards, 1985; Dries, 1989; Lema, 1996). Increasingly, however, tropical wetlands are under threat from developers and planners motivated by economic gain, who fail to appreciate the range of functions and services wetlands provide, in particular for local communities whose subsistence is often linked to the wetland ecosystem (Hollis, 1990; Turner, 1991). Even at the local level, the need for more agricultural land and greater farming productivity has placed wetlands under threat within local communities. Throughout many parts of highland East Africa in particular, the community drainage of wetlands is regarded as a means of increasing arable land and ultimately food production (Denny, 1993a).

There is now, however, a growing awareness that wetlands provide a variety of functions which include not only the provision of wetland products which can be exploited but also critical ecohydrological roles which, according to Barbier (1993), provide 'environmental services' for many local communities. Most of the small-scale economic activities associated with wetlands depend upon these environmental services.

Discussions of the functions associated with tropical wetlands are numerous (Adamus and Stockwell, 1983; Maltby, 1986; Dugan, 1990, Roggeri, 1998) and considerable research has been carried out on the specific roles wetlands play and how these interact with the local environment. The following sections present an overview of the key ecohydrological and socio-economic functions of wetlands reported in the literature, which are of particular importance to rural communities throughout the developing world.

Environmental functions

Depending upon the nature of their ecohydrological relationship, each wetland is able to perform a variety of functions which play a key role in the regulation and stability of the physical environment. These include:

Water table recharge and discharge When the velocity of water entering a wetland is reduced and its subsequent residence time in the wetland increases, there may be some percolation of the water downwards into the aquifer and consequently water table recharge occurs (Mihayo, 1993). Infiltration also occurs during floods when the velocity of overland flow is reduced and the inundated land surface is increased.

As a result of their lowland position in relation to surrounding land many wetlands also act as sinks for water discharged from aquifers (Roggeri, 1998). The relationship between groundwater and wetlands is extremely complex and dependent on many factors such as regional groundwater flow systems, geology, hydraulic conductivity, slope and relief of the catchment (Carter and Novitski, 1988). This means there is considerable variation between different types of wetlands in their ability to perform a groundwater recharge / discharge function. Even if a wetland has the ability to perform such a regulatory function this may be variable and seasonal (Siegel, 1988).

Flood control and river regulation According to Balek (1983), the volume of run off in a catchment will decrease as the percentage area of wetlands increases. Wetlands are able to mitigate floods by storing potential floodwaters, reducing floodwater peaks and ensuring that floodwater from tributaries does not all reach the main river at the same time (Maltby, 1986). Krhoda (1992) also suggests that the presence of wetlands in a catchment may influence run off by having high evapotranspiration rates. Flood regulation may be further affected by the actual slope of the wetland which may promote infiltration and consequently water table recharge. During the dry season, subsurface flow from saturated wetlands may replenish stream flow.

Research by Amatya *et al.* (1995) suggests that the position of a wetland in a catchment is the critical factor in regulating stream flow. Small wetlands (5 per cent of catchment area) located downstream were reported as being just as effective as large wetlands (40 per cent of catchment area) upstream. Wetlands as a water storage facility were found to be significantly more effective at reducing annual outflow from the catchment than agricultural land.

Sediment trapping As the velocity of water decreases on entering a wetland as a result of flooding, the presence of vegetation or an increase in channel width, suspended sediment settles in the wetland. The destruction of wetlands can seriously affect this process and lead to downstream sedimentation. For example, the Tana River in Kenya has shifted its channel in its lower reaches as a result of sedimentation caused by the drainage of the Yala swamp upstream. This has resulted in the diversion of the river channel and flooding which has had a dramatic effect on the economic activities associated with this area (Krhoda, 1992).

Maintenance of water quality The practice of discharging wastewater into natural wetlands has been used as a means of waste disposal for hundreds of years (McEldowney *et al.*, 1993). Research on the ability of wetlands to purify water has shown that the anaerobic conditions which exist within wetlands, enhances the retention of many compounds and facilitates processes such as denitrification, ammonification and the formation of insoluble phosphorous-metal complexes (Bastian and Benforado, 1988). Wetland vegetation such as *Eichornia* is also able to store large quantities of nutrients and heavy metals (Gopal, 1987).

Biosphere and micro-climate stabilization The conditions of high humidity and evapotranspiration found in many wetlands, may significantly affect local and regional climates (Roggeri, 1998). In addition, the process of microbial decomposition is encouraged in wetland ecosystems and this can lead to storage or emissions of gaseous by-products which may be important for global atmospheric stability (Odum, 1979). Peatlands are well documented as 'sinks' for atmospheric carbon although drainage and destruction leads to the rapid loss of stored carbon as decomposition occurs (Immirzi *et al.*, 1992).

Wildlife habitats Wetlands are host to a rich biodiversity, offering a range of ecological niches for wildlife both spatially and temporally (Maltby, 1986; Denny, 1994). Furthermore, they often represent areas of high endemicity for rare or endangered species (Dugan 1990). In seasonally inundated wetlands different species have adapted to conditions during the dry season and the wet season whereas in permanent wetlands, species may have evolved in ecological isolation and represent an endemic and rare population (Turner, 1988). Dugan (1993) presents a variety of specialized plant adaptations to wetland environments which include among others *Sphagnum* spp., which is tolerant of the extreme acidic conditions found in some marshes, and a range of aquatic and emergent plants including *Cyperus papyrus*, *Pistia stratiotes* (water lettuce) and *Eichhornia crassipes* (water hyacinth). In addition, many areas of wetland support high concentrations of endemic fauna. For example, the Bangweulu basin in Zambia provides a habitat for 30,000 black lechwe antelope (*Kobus lece smithemani*) and it constitutes one of Africa's most important areas for sitatunga (*Tragelaphus spekei*). Wetlands provide vital habitats for waterfowl and migrating birds, the main factor which served as the impetus for the Ramsar convention (Matthews, 1993).

Socio-economic functions

The socio-economic functions provided by tropical wetlands tend to be associated with the direct exploitation of wetland products for economic gain or subsistence. The most ubiquitous form of direct wetland use is wetland agriculture which in many developing countries has been carried out in a small-scale, low input, sustainable manner for thousands of years (Scoones, 1991; Reij *et al.*, 1996).

Agricultural production Farming activities are major economic pursuits around many wetlands in the tropics where crops such as rice, maize, and various vegetables and fruit are cultivated (Dries, 1989; Soerjani, 1992; Omari, 1993). Seasonally inundated floodplains such as those in West Africa, are economically important farming resources. Farming methods have been developed which maximize the use of these areas throughout the seasons, both during the flood period and especially after it has fallen (Adams, 1993a). The advantage of using floodplain soils is related to the ecohydrological characteristics mentioned above. They are relatively fertile areas, being inundated with sediment from the catchment and the presence of clay soils ensure a long period of water retention.

Indigenous knowledge systems (IKS), discussed further in Chapter Three, have been identified as playing a major role in the way wetlands are used for cultivation. To adapt to the annual flooding of the Zambezi, the Lozi of the upper Zambezi floodplain have evolved their own unique traditional methods of wetland cultivation to ensure the sustainable use of wetland resources (Wood, 1985; Chiuta, 1995). These have included the allocation of different cultivation systems and soil conservation measures to hydrologically different areas within the floodplain (Thole and Dodman, 1996). Agricultural production in wetland areas is frequently characterized by adaptations to the uncertainty of flooding, such as the construction of flood defences and small-scale irrigation initiatives (Kimmage and Adams, 1992).

Plant production Natural wetland plants also serve a variety of functions. Soerjani (1992) points out that of the 266 species of weeds associated with wetland rice cultivation in Indonesia, 70 per cent can be utilized in a range of activities including medicine, cattle fodder, household purposes and for human consumption. Zerihun Woldu (1998) reports that in the valley bottom wetlands of south-west Ethiopia, various wetland plants are collected for medicinal purposes ranging from headaches to rabies treatment. On Lake Tana in Ethiopia, locally harvested papyrus has been used in the construction of fishing boats for hundreds of years (Muthuri, 1993). Throughout East Africa, papyrus and other reeds are also used as a construction material for roofing (Denny, 1993b).

Fishing Fish production is a basic element in the economy of many tropical wetlands. In the developing world there is often a localized economic and nutritional dependence on this resource (Maltby, 1986; DeMerona, 1992; Bwathondi and Mwamsojo, 1993).

Dry season cattle grazing Seasonal wetlands can provide a valuable resource for livestock grazing as a result of the high biomass associated with these areas. In many of Africa's savannahs where the climate is semi-arid and rainfall is seasonal, wetland grazing is widespread (Scoones, 1991), with wetland landforms such as the *dambos* of Zimbabwe and *fadama* of Nigeria being important seasonal grazing resources (Roberts, 1988; Turner, 1994).

Water supply Most wetlands can provide a potable supply of water for the surrounding population (either directly or from springs which contribute to their formation) which is a critical function in many semi-arid or seasonally dry areas (Scoones, 1991). Depending upon their ecohydrological characteristics, many wetlands are able to purify water supplies as a result of the effects of microbial action. In Sudan, Jahn (1981) describes how local communities associate the growth of *Cyperus latifolius* with clean drinking water. A wetland's ability to regulate and store water can also be beneficial in the production of hydro-electric power by improving the supply of water for power production.

Socio-cultural functions

In many developing countries wetlands also have a value which goes beyond the direct utilization of products and the benefits of supportive hydrological functions. Barbier (1993) describes this as 'non-use value', and in developing countries this can include the cultural or religious importance attached by local communities to wetlands. For example Denny (1994) cites African tribes which consider the clay from particular wetlands as sacred, using it to smear over their bodies during circumcision ceremonies.

Another important non-use benefit is the aesthetic value of wetlands which is often exploited for tourism. Whilst wetland tourism is well developed in many developed countries, the tourism potential of wetland sites has only recently been realized in developing countries. Maltby (1986) suggests that in Botswana alone, wildlife safaris are worth over $15 million annually. Throughout East Africa there is a growing recognition of the importance of wetlands as major wildlife habitats, which offer significant potential for tourism. In Zimbabwe and Zambia in particular, wetland tourism is being developed as a component of a wider rural development programme, in that local communities are given the responsibility of managing wetlands for their aesthetic and in return they receive economic and social benefits from tourism (Chabwela, 1992b; Sanyanga, 1994).

In summary, wetlands have the potential to perform a wide range of environmental regulatory functions, sustain a variety of socio-economic activities and in many areas, fulfil a spiritual role. This is clearly demonstrated within the African continent where they play a critical role, either directly or indirectly, in sustaining a large proportion of the population (Drijver and Marchand, 1985; Scoones, 1991; Adams, 1992; Acreman and Hollis, 1996).

Wetland types and their utilization in Africa

Wetlands in Africa account for approximately one per cent of the total land area (Hughes 1996; Thompson, 1996), although this depends upon the criteria used to identify wetlands. Thompson (1996) points out that this figure underestimates the total area of seasonally inundated grasslands across the continent and it does not include the various deep lakes found throughout East Africa. The variety of wetland types found within Africa reflect the variation in climate and geomorphology. Denny (1993a) also suggests that Africa is composed of two physiographic units which influence the occurrence of wetlands. Low Africa, situated to the north and west, is composed of sedimentary basins and upland plains below 600 m a.s.l., which favour the formation of floodplain wetlands. In contrast, high Africa to the south and east, is generally characterized by the results of tectonic activity which include extensive mountainous areas, deep valleys and highland plateaux. Within this higher area lakes and swamps tend to be more abundant.

Among the variety of wetland types found within Africa seasonally inundated floodplains and swamps constitute the most extensive, accounting for almost half of the continent's total wetland area (Drijver and Marchand, 1985). Research has suggested that these areas are critical resources for Africa's human population, especially in areas where rainfall is seasonal and wetlands represent a reliable source of water throughout the year (Hollis, 1990). The following sections highlight the ways in which three types of wetland, each abundant in Africa, provide a range of functions that support a variety of human subsistence and economic strategies.

Freshwater swamps

The term 'swamp' within the context of tropical Africa usually refers to a wetland distinguished by its emergent herbaceous vegetation, which is predominantly composed of either reeds (*Phragmites*), papyrus sedges (*Cyperus papyrus*) or bulrush (*Typha*) (Denny 1993a; Roggeri, 1998) although vegetation varies according to specific hydrological characteristics (Beadle, 1981). Hydrologically a swamp is characterized by land which is seasonally or permanently flooded with shallow water originating from groundwater, springs, streams or runoff (OECD, 1996) depending upon the geomorphological setting. Kimble (1960) suggests that African swamps share several key characteristics:

1. they are runoff regulating systems acting as reservoirs with an increased rate of evapotranspiration;
2. they have a high ratio of surface area to water depth;
3. the size of swamps fluctuates from year to year and from season to season;
4. they show three clearly marked zones: at the margin of the swamp is a zone under water for only a brief part of the year, a second zone is waterlogged for a much longer period and a third zone is under water throughout the year.

Balek (1977) estimates the total size of tropical swamps in Africa at 340,000 km^2, which, according to Beadle (1981), possibly exceeds the area of all the open waters of all African lakes. The distribution of wetlands exhibiting swamp characteristics is widespread. The largest continuous area of swamp in Africa is located in the Sudd region of Sudan at the confluence of the Nile and Bahr-el Ghazal rivers (Hughes, 1996). The area is characterized by approximately 40 000 km^2 of permanent papyrus swamp and when the inflow rivers are in flood, the total area of inundation almost doubles (Beadle, 1981). Seasonally inundated areas of the Sudd provide rich vegetation on which cattle are grazed during the dry season, supporting nomadic and semi-nomadic communities. The areas of permanent papyrus swamp also provide a vast habitat for wildlife which consequently support hunting and fishing activities (Dugan, 1993). In contrast, small swamps with steep sided valleys are widespread throughout the East African highlands, in particular Rwanda, Uganda (Beadle, 1981, Denny and Turyatunga, 1992; Denny, 1993b), southern Tanzania (Lema, 1996) and Ethiopia (Balek, 1977, Bayessa Urgessa, 1995).

These wetlands represent significant resources to local communities. For example, there is evidence to suggest that small upland valley swamps have traditionally been utilized for shifting, seasonal cultivation which, according to Denny (1993a), does little damage to the overall structure of the swamp and provides food for the local population. Lema (1996) suggests that the drainage and cultivation of valley bottom swamps in southern Tanzania has been carried out since the 19th century. These areas represent an important agricultural and water resource to local communities who, during the dry season, cultivate a range of crops on raised beds known as *vinyungu*. The system of wetland drainage and cultivation is well established and the characteristics of both drainage channels and *vinyungu* are adjusted each year depending upon rainfall, to ensure a successful crop. With an increasing local population placing more demands on these areas, however, there is some doubt as to whether they can continue to sustain their current level of agricultural use.

Small valley swamps are also abundant throughout many parts of West Africa where they provide a range of benefits for local communities (Dries, 1989, 1991; Richards 1985). Richards (1985) describes how the majority of farm households in Sierra Leone engage in a well developed system of rice cultivation in inland valley bottom swamps. In addition to rice, seasonally flooded swamps are also cultivated during the dry season with crops such as wheat, tobacco, onions and tomatoes. According to Dries (1991) this system of inland swamp cultivation became prominent throughout Sierra Leone following the harvest failure of 1919, when farmers were forced to move rice production into marginal areas. Since then, most farmers in addition to upland farming, cultivate valley bottom swamps which offer significant advantages in soil fertility over upland soils. During the 1970s, however, several agricultural development programmes were initiated to intensify wetland rice production in response to population pressure and increasing demands for rice. Dries (1991) suggests, however, that these initiatives have remained largely unsuccessful compared to the farmers' own indigenous rice cultivation techniques.

Denny (1993b) suggests, however, that the natural functions and indigenous utilization strategies are under threat from the continued encroachment of local communities and from large scale schemes which influence the wider catchment hydrology. The problems associated with the intensification of local wetland use is clearly illustrated by Denny and Turyatunga (1992) who describe how the small, upland valley swamps found throughout Uganda have undergone intensive drainage to increase agricultural production on the wetlands' deep organic soils. Traditionally these areas were used as a source of livestock fodder, craftwork material and water supply and, in addition, there has been some limited cultivation of the wetland margins during the dry season (Tukahirwa, 1989). Government sponsored drainage was initiated in 1955 on many upland swamps and although crop production was initially high, certain environmental problems shortly began to develop. In particular, local communities suffered from a shortage of reeds which they used as a raw craft material, spring and well water became less plentiful and flooding of the swamp became more unpredictable and damaging. Denny and Turyatunga (1992) point out that these problems have stemmed from the impact of drainage on the hydrological regime of the swamp. Whereas in a natural state the swamp acts as a sponge which regulates the flow of water from the catchment, under drained conditions water is conveyed rapidly downstream reducing the swamp's ability to buffer peak flows and retain nutrients. In addition, shrinkage of the soil and the oxidation of organic matter has also been reported. The overall effect has been a degradation of the wetland resource to the point where the wetland's only utility lies in pasture and, according to Denny and Turyatunga (1992), even this is dominated by non-palatable vegetation such as *Cyperus latifolius*.

Similarly Gichuki (1992) suggests that valley bottom swamplands throughout Kenya are also under threat from agricultural intensification which has been stimulated by increasing population pressure. Whilst the intensification of agriculture in the valley bottoms has led to an increase in crop yields, it has also had negative environmental and socio-economic impacts, similar to those reported in Uganda (Denny and Turyatunga, 1992). These include a decrease in the biodiversity of wetland plants which may be of specific value to local communities, soil erosion and the degradation of the soil resource and critically, the reduced availability of water for human or livestock consumption.

Although Africa's freshwater swamps are increasingly being recognized as important natural resources, there has been a tendency for international agencies and researchers to focus on larger areas located within floodplains or around lake margins which are significant in terms of their wildlife resources. Smaller areas of swampland such as those in Uganda, have received less attention and there is a need to understand the functioning of these areas from an interdisciplinary perspective if their sustainable utilization is to be achieved (Denny and Turyatunga, 1992).

Riverine floodplains

Riverine floodplains are alluvial lowland areas which are located adjacent to rivers and which become seasonally flooded as a result of variations in the river discharge (Roggeri, 1998). Their occurrence in Africa is associated with long rivers flowing through areas of high, seasonal rainfall, although seasonal flooding in the major floodplains of West Africa is dependant on rainfall in the highlands of southern Guinea and Sudan zones (Thompson, 1996). The situation is complicated in the case of longer rivers such as the Niger, when different parts of the river catchment contribute runoff at different times of the year and the result is two distinct periods of flooding. Adams (1993a) also makes the distinction between the upper reaches of a river, which experience a more peaked hydrograph, and the lower reaches, which are wider and which subsequently experience a slower rise in floodwater. In addition, the wide variety in geomorphological characteristics influenced by fluvial processes also influence the spatial and temporal characteristics of floodplain inundation. This has a profound effect on vegetation which ranges from that typical of ox-bow lakes and herbaceous marshes to that of a swamp forest wetland environment (Hughes, 1996).

The ecohydrological functions linked to periods of inundation include aquifer recharge, flood control and sedimentation, while socio-economically, floodplains have immense value and utility to human populations in terms of their foraging, grazing, hunting, fishing and agricultural resources (Adams, 1993a). Floodplains may be able to perform a range of functions and provide numerous products simultaneously, or serve different communities in different ways throughout a year (Adams, 1996). In particular, floodplain agriculture, which involves cultivation on the rising and falling floodwaters, is one of the most extensive forms of indigenous irrigation in Africa. Floodplain soils, which often contain clay, characteristically retain moisture long after flooding has taken place and farmers have built up a detailed knowledge of both the ecological requirements of certain crops and the variation in soil moisture availability within particular floodplain wetlands (Gluckman, 1968; Adams, 1996). McIntosh (1983) suggests that the inner Niger delta was the location of prehistoric experimentation with the domestication of rice, which according to Richards (1985) remains an on-going process in West Africa's floodplain wetlands.

The Hadejia-jama're floodplain in Nigeria is typical of the significance of floodplain wetlands as grazing resources. Here sheep, goats, cattle and camels are grazed by both nomadic and sedentary farmers who recognize the importance of the rich, seasonally exposed grassland (Adams 1993b). Although traditional grazing rights exist for the Fulani pastoralists, they have recently come into conflict with farmers eager to exploit the residual moisture and high water table in the floodplain grasslands. Increasing agricultural exploitation has also occurred as small-scale irrigation technology has been widely adopted, which in many areas has included the use of petrol driven irrigation pumps (Kimmage and Adams, 1992). In addition, grazing areas have been degraded as a result of the harvesting of hay which is exported to local markets for horse feed (Adams, 1993a).

Similarly in the Kafue flats floodplain of Zambia, utilization of the floodplain has become unsustainable as four different ethnic groups compete for grazing land, agricultural land and fishing resources (Chabwela, 1992a). In recent years, fisheries have declined as a result of over-exploitation fuelled by increasing population pressure and the lack of catch management practices (Jeffery et al., 1992).

While over-exploitation represents a significant threat to floodplain and swamp utilization, the riverine nature of floodplains also makes them targets for large scale water resource development projects, in particular dam construction. Research has suggested that dams have had serious adverse environmental impacts on floodplain wetlands in terms of the disruption of seasonal flood regimes (Adams and Hughes, 1990). According to Masundire (1996) the major impact of the Kariba dam on the floodplain of the lower Zambezi has been the absence of flooding and consequently a state of nutrient impoverishment. Flows are also linked to the demand for hydro-electric power which in turn fluctuate and contribute to the unpredictability of flooding. In the Hadejia-Nguru floodplain of Nigeria, the construction of the Tiga dam and its impact on the flooding regime has increased the pressure on a resource base which has already reached its carrying capacity in terms of human exploitation (Thomas, 1996).

Dambos

Much research has been undertaken on the ecohydrological characteristics and the human utilization of *dambos*, the common name given to a form of seasonally inundated swamp grassland which occur on the plains of eastern, southern and western Africa. The term *dambo* is of Zambian origin and refers to a specific type of intermittent swamp which has similarities with the *vleis* of South Africa, *fadamas* of Nigeria and *bolis* of Sierra Leone (Balek, 1977; Turner, 1986). A *dambo* can be defined as:

> ...periodically inundated grass-covered depressions on the headwater end of a drainage system in a region of dry forest or bush vegetation (Ackermann, 1936, p.150).

Mackel (1985) states that *dambos* are the result of a particular geomorphological form and vegetation type based on the seasonally wet and dry climate and edaphic conditions. Typically, *dambos* are zero order basins characterized by a shallow linear depression. They may vary in width from a few metres to over 1 km (Turner, 1986) and exhibit a variety of shapes in headwater areas, although they become more linear downstream (Acres *et al.*, 1985). Balek and Perry (1973) argue that *dambos* differ from swamps in that the main source of water recharge is direct rainfall whereas other types of swamp commonly receive water input from stream channels fed by surface and sub-surface runoff from the catchment. Bullock (1992b) in a reassessment of *dambo* literature, however, suggests that overland flow, subsurface throughflow and overbank contributions from river channels are, in addition to rainfall, significant water sources. Several definitions cited by Whitlow (1985) also draw attention to *dambos* occurring in land which is gently sloping, in contrast to other types of swamp which are abundant in highly dissected terrain.

The seasonal waterlogging of the *dambo* produces distinctly different vegetation to that of the surrounding land which is most frequently characterized by *miombo* woodland (Roberts and Lambert, 1990). The *dambo* itself supports a variety of plant communities which reflect the different hydrological conditions within it. The *dambo* margin is dominated by grasses and as it gradually becomes wetter, these give way to sedges and herbaceous plants. The role of vegetation in the hydrological functioning of *dambos* is, however, ambiguous (Bullock, 1992b). Whilst some authors suggest that vegetation impedes drainage and aids water storage (Balek and Perry, 1973; Whitlow, 1983), others argue that it produces high evapotranspiration losses thereby inducing a depletion in water storage and baseflow (Smith-Carrington, 1983).

Central to most studies on *dambos* and their similar landforms has been the assertion first proposed by Balek and Perry (1973), that the *dambo* acts as a reservoir, storing water during the wet season and releasing it slowly during the dry season, maintaining a stream baseflow. This 'sponge theory' is, however, a contentious issue and, according to Faulkner and Lambert (1991), an oversimplification of the physical processes within *dambos*. Their research on *dambos* in Zimbabwe suggested that the storage capacity of the catchment area above the dambo is influential in maintaining a baseflow of water throughout the year. Bullock (1992a) suggests that the ability to maintain dry season flows depends upon the specific soil characteristics of the area in addition to catchment land-use practices and the ways in which *dambos* themselves are utilized.

The availability of soil moisture in *dambos* during dry periods remains the principal reason behind their agricultural utilization, either as areas for crop cultivation or cattle grazing (Turner, 1986). Even at the end of the dry season some parts of *dambos* will remain wet, with a water table level between 2 and 0.5 metres below the soil surface. This permits the growth of natural vegetation which can be used for grazing, or allows crop cultivation at a time when there are usually food shortages.

Small-scale agriculture in *dambos* is usually practised in the form of maize and vegetable 'garden' cultivation, based on indigenous water management techniques. During the dry season, production is aided by the use of 'micro-scale irrigation' methods which offer the potential for improved food security and income, without high investment costs (Lambert *et al.*, 1990). Dry season cultivation also has the advantage of being less susceptible to pests and diseases (Turner, 1986). During the rainy season when most of the *dambo* remains flooded, rice can be grown whilst some dry season crops can be cultivated on the *dambo* margin. In climatically drier areas maize cultivation can also be attempted during the wet season without irrigation (Roberts, 1988). As well as having available water, *dambos* offer a relatively high soil fertility in contrast to that of the surrounding drylands. The decomposition of the grass and sedge dominated vegetation leads to the build up of organic matter in the *dambo* soil, which in some cases is similar to peat.

Dry season cattle and livestock grazing is another important form of *dambo* utilization (Roberts, 1988). The *dambo* can produce a high biomass which is available during the dry season when upland grazing resources are exhausted, although the quality of this is related to the nutrient status of the soil. Acres *et al.*, (1985) suggest that a *dambo*, used seasonally for grazing, can have a carrying capacity of one livestock unit (250 kg) for between 0.5 and 5 ha depending on the quality of the grazing vegetation. Since *dambos* represent an attractive resource for both grazing and agricultural production, there is often a conflict of demands on these areas. In Nigeria, *fadama* are an important traditional grazing resource for Fulani pastoralists but increasingly these lands have been the subject of an extension of agriculture, instigated by the government and development agencies (Turner, 1986; Binns, 1994). Similarly in Tanzania, wheat farming in the traditional valley bottom grazing areas of the *Barabaig* pastoralists, has disrupted their complex system of rotational grazing. This has resulted in a loss in both livestock productivity and their associated economic benefits (Scoones, 1991). Competition for these wetland resources has inevitably led to conflict and the degradation of *dambo* wetlands in terms of their ability to continue to supply a range of functions.

Research has suggested that the degradation of *dambos* is caused by the intensification of their use, in particular both grazing and cultivation, whose effects are exacerbated by the influence of low rainfall (Roberts, 1988). Intensification tends to occur in those *dambos* located in arid areas where there is a heightened need for their hydrology dominated functions, making them particularly fragile resources (Whitlow, 1985; Turner, 1986). *Dambo* degradation, as a result of either over-grazing or intensive cultivation, is characterized by:

1. a lowering of the water table;
2. a reduction of surface vegetation leading to exposure of bare soil;
3. increased evapotranspiration;
4. erosion and gully formation;
5. the reduction of organic matter and loss of fertility;
6. the desiccation of soils (Roberts, 1988; McFarlane and Whitlow, 1990).

The overgrazing of livestock is seen as the major factor contributing to soil erosion and gully formation in the *dambo*, whereas the main effects of intensive cultivation are a decrease in fertility and a fall in water table level (Roberts, 1988; McFarlane and Whitlow, 1990). The effects of over-cultivation are made worse by the use of small-scale irrigation schemes and drainage ditches which increase surface runoff and lower the water table (Turner, 1986).

The processes behind the ecohydrological functioning and degradation of *dambos* remain ambiguous, despite the attention they have received. Research has suggested that *dambos* contribute to downstream water flows as a result of their water storage properties (Balek and Perry, 1972; Acres *et al.*, 1985) but evidence also suggests that *dambos* cause a reduction in streamflow as a result of high evapotranspiration rates from their surface and catchment (Bullock, 1992a). As Scoones (1991) points out, it is difficult to predict the impact of agriculture on a *dambo* or the value of the *dambo* for catchment hydrological processes without very site specific information. This is necessary before any ecohydrological assessment or sustainable utilization study can be made.

Wetland management options

The underlying theme from this review of swamps, floodplains and *dambos* in tropical Africa is that they are important natural resources, particularly for local communities whose livelihoods are often dependent on them, yet they are increasingly under threat from development activities. Highland swamps are threatened by the intensification of drainage for agriculture, *dambos* by intensive cultivation and over-grazing, and floodplains by these activities combined with the effects of large water resources development initiatives. The functions and benefits of wetlands are, however, slowly being recognized by officials and researchers, and in many parts of Africa, wetland conservation or environmentally sensitive management has become a priority (Maltby *et al.*, 1992; Hollis *et al.*, 1993; Matiza and Crafter, 1994). In particular, methodologies for identifying the functions of wetlands have been established (Adamus and Stockwell, 1983; Brinson *et al.*, 1994, Maltby *et al.*, 1994) and recently the economic assessment of wetland functions and benefits has legitimized conservation and sustainable management above development (Turner, 1991; Barbier, 1993, Barbier *et al.*, 1997).

Whereas at one time the Ramsar Bureau advocated conservation of wetlands primarily for wildlife, an appreciation of the wider importance of wetlands, particularly for human communities, has stimulated holistic appraisals of wetlands and resulted in a new range of management recommendations (Denny; 1994; Roggeri, 1998).

Wetland conversion

Although it is now argued that wetland conversion is considered to be synonymous with a complete loss of natural wetland functions and benefits (Maltby and Turner, 1983; Hollis, 1990; Dugan, 1990), the conversion of wetlands to agricultural land remains an attractive option in many parts of the developing world where achieving a state of food security is often a key issue. Roggeri (1998) argues that the conversion of wetlands to a different ecosystem (e.g. via drainage) can be justified in specific circumstances, in particular where:

1. social and economic needs are particularly pressing;
2. wetland conversion is the only solution which would meet these needs;
3. it has been demonstrated that the wetland could not contribute significantly to meet these needs;
4. it has been demonstrated that the wetland is of minor value (and that its value will not increase in the future) (Roggeri, 1998, p.86).

In addition, Dugan (1992) suggests that the conversion of wetlands should be considered where an assessment of the costs and benefits involved show that conversion is the most effective means of contributing to the well being of the human population.

Wetland conservation

At the opposite end of the scale, wetland conservation is the only management option where the wetland is so rich that any alternative management strategy has a detrimental impact (Dugan, 1992). 'Richness', however, is a highly subjective term. From an economic perspective it can refer to the diversity of wetland functions, which should be conserved where the wetland contributes significantly to the local or national economy (Roggeri, 1998). Alternatively it can be interpreted from a biodiversity perspective whereby any reduction in the gene pool can be regarded as degradation (Denny, 1994). In a consideration of the management options for the Hadejia-Nguru floodplain in Nigeria, Adams (1993c) argues that conservation alone is 'profoundly unfair to local people', although this depends on the degree to which human utilization of wetland resources is already taking place.

In terms of ecological richness or biodiversity, Denny (1994) suggests that wetland conservation can be justified on the grounds that human knowledge is insufficient to make any assessment of biodiversity and at which stage it becomes irretrievably damaged. If this is the case, wetlands should be conserved or used on a wholly sustainable basis to ensure no loss to biodiversity or their natural functions. Furthermore, the wetland may provide a valuable gene pool which is of benefit to the human population, e.g. medicinal plants and crop varieties. The risk of losing this resource is considered too great if wetland development proceeds.

Sustainable wetland management

Somewhere between wetland conversion and conservation lies the integration of both, where management maintains the functions and benefits of wetlands, while also sustaining a certain level of utilization so that the value of wetlands to the local community is enhanced. According to Roggeri (1998) such forms of management may involve local communities continuing their production activities or carefully planned micro-development initiatives aimed at increasing productivity for economic benefit. It is important that utilization does not exceed the regeneration capacity of the natural resources or result in any changes in the functioning of the wetland as a whole. Dugan (1992) suggests, however, that the majority of wetlands will be able to undergo some degree of exploitative activity which can provide sustainable benefits.

The alternative to limited exploitation according to Roggeri (1998) is the complete conservation of one area within a wetland and the complete development of another, hence the preservation of wetland functions alongside the creation of new economic opportunities. Such a strategy, however, could only be justified if, again, there is no net loss of original wetland ecosystem of functions and where development does occur, it is carried out on a permanent and sustainable basis.

The origins of this sustainable management approach to wetlands date back to the third meeting of the Ramsar Convention Bureau in 1987 which developed the concept of the 'wise use of wetlands'. The increasing number of developing countries ratifying the convention necessitated a shift in emphasis from management for wildlife to wetland management for people's subsistence. 'Wise use' was defined as:

> ...their sustainable utilization for the benefit of mankind in a way compatible with the maintenance of the natural properties of the ecosystem (Matthews, 1993, p.54).

In addition, sustainable utilization was defined as:

> Human use of a wetland so that it may yield the greatest continuous benefit to present generations while maintaining its potential to meet the needs and aspirations of future generations (Matthews, 1993, p.54).

In order to achieve sustainable utilization, the Ramsar Bureau also presented several guidelines to be followed by governments and planners. The first involves carrying out an inventory of wetlands in order to identify the range of wetland resources within a specific area. Once this has been achieved, an evaluation of the functions and values of each wetland type should be carried out so that their importance in the landscape, whether this be socio-economic or biodiversity, can be established. With this information a wetland's management potential can be ascertained, although, prior to any intervention in the wetland, an Environmental Impact Assessment (EIA) should be undertaken and continued throughout the course of any development initiative.

Uganda provides an example of where these guidelines have been followed with some success. A national wetlands' policy has been developed which aims to sustain the ecological and socio-economic functions of wetlands for the future well being of the population (Howard, 1991; Mafabi and Taylor, 1993). The policy is based on an understanding of the Ramsar convention (which Uganda ratified in 1988), while also recognizing the importance of wetlands to rural communities. Consultations with District Development Committees, which include people at all levels of the community, aim to increase public awareness of the importance of wetlands, but also ensure that their own attitude towards wetlands contribute to realistic and practical policies (Lwanga, 1999). The result has been the development of integrated farming practices and sustainable production activities in many of Uganda's wetlands (Government of the Republic of Uganda, 1994).

Whilst most would agree that the Ramsar guidelines provide a good starting point for sustainable wetland management planning (Hollis et al., 1988; Adams, 1993a; Roggeri, 1998), there is some concern that small, less conspicuous wetlands can remain unobserved (Chambers, 1990; Scoones, 1991). In many areas these small wetlands are more significant especially in terms of their importance for local communities (Scoones, 1991). In terms of wetland functional assessment techniques, smaller wetlands which may perform a small number of functions are likely to be valued less highly than those with a greater number and in this respect they are less likely to be considered important by wetland planners. Furthermore, as Pearce and Turner (1991) point out, such sites are more likely to undergo development because of a lack of information regarding their importance. Hollis (1990) also suggests that smaller wetlands are less likely to undergo any form of EIA for both financial and practical reasons.

The Ramsar Convention and, to some extent, the majority of wetlands research over the past 30 years have been typified by an approach which is biased towards large, multi-functional, prestigious wetlands. The Ramsar Convention designates wetlands of international importance on the basis of their wildlife populations, hence those with less conspicuous wildlife are considered less important. Furthermore, only those countries which are signatories to the convention are compelled to follow any wise use or sustainable utilization principles in any of their wetlands. For many developing countries, ratifying the Ramsar Convention is not an option given the dependence of local communities on wetland resources.

'Traditional' wetland management

While it would appear that small wetlands are becoming increasingly threatened as a result of a lack of management intervention, there is some evidence to suggest that these areas have avoided destruction because of their indigenous management or lack of intervention by government or external agencies. Scoones (1991) argues that the expansion of agriculture into wetland areas throughout the developing world has been a way of spreading the risks of farming for generations and, when used in traditional ways, this indigenous management of wetland resources is sustainable.

In West Africa, both Richards (1985) and Adams (1993a) provide clear examples of these indigenous wetland management techniques in operation. Turner (1994) in a review of small-scale irrigation in developing countries, also demonstrates how farmers' use of wetlands is well adapted to their environment and how any management intervention from outside the community can have significant negative impacts, resulting in a shift to unsustainable use.

The main implications here are first, that wetland utilization can be sustainable without any outside intervention and secondly, where wetlands are under threat, local communities may have the capacity to solve their own problems. For those smaller wetlands that are under threat from a variety of socio-economic or environmental pressures, the empowerment of people to manage these indigenous resources potentially offers a means of attaining sustainable management with the minimum of outside intervention. As discussed in the following chapter, a recognition of the significance of indigenous knowledge and its role in achieving sustainable natural resource management has been at the forefront of rural development since the 1970s. Only recently, however, have wetland users been considered important in wetland management and there is still little evidence to suggest that their knowledge and skills are being empowered in the planning process.

Chapter Three takes this discussion forward, highlighting the significance of indigenous knowledge in the development process and the implications of this for sustainable natural resource and wetland management.

Chapter 3

Indigenous Knowledge and Wetland Management

Introduction

Following the recognition in Chapter Two that wetlands are an important resource to human populations, this chapter discusses the role of indigenous knowledge (IK) in traditional natural resource management strategies and in particular, wetland utilization. IK is local knowledge, that is knowledge which is unique to a given culture or society and increasingly it is being regarded as an important resource which can facilitate development in participatory and sustainable ways (Warren, 1991). This chapter addresses some of the debates surrounding the recent upsurge in interest in IK, the characteristics of IK, its utilization in the development process and its prominent position on the international development agenda at the present time. Finally, the extent to which the IK and development paradigm can be transferred and applied to the field of wetland management is examined.

Natural resource development in the developing world

The 1960s, which saw much of Africa gaining independence from colonial powers, were characterized by widespread demographic, economic and political changes, resulting in poverty, food shortages and environmental problems. The need for development, a process which attempts to improve the living conditions of people (Bartelmus, 1986), stemmed from a recognition of the problems which occurred throughout the developing world during this period. Economic pressures linked to the debt crisis and the downturn of the world economy during the 1980s, have further increased the need for production and economic growth. As a result, the accelerated development of the natural resource base, in particular the agricultural sector, has become the objective of the world's governments and the UN in an attempt to alleviate poverty in the developing world (Adams, 1990; Sachs, 1992).

In this context, development has been synonymous with modernization and industrialization, whereby developing countries mimic the models of economic growth which have been 'successful' in the developed world (Sachs 1992; Reid, 1995). In particular, the development and modernization of agriculture in developing countries, which has existed at the subsistence level for hundreds of years, has been viewed as a key to economic growth both at the community level and for the country. Central to such agricultural development initiatives has been the transfer of technology (TOT) approach, which has also been regarded as 'scientific' or 'western' style development (Howes and Chambers, 1979). The green revolution, which occurred throughout the UN development decades of the 1960s and the 1970s, typified this approach to development as it sought to transfer what Reijntjes *et al.* (1992) term high external-input agriculture (HEIA) to much of the developing world.

The green revolution came about through the application of science and technology to the problems of poverty and food shortages in the developing world. According to Dixon (1990) it was characterized by:

1. a breakthrough in plant breeding that produced high yielding food grains;
2. a package of technology, including fertilizer, insecticide, implements, water control and high yield seed varieties;
3. an agricultural development strategy in which the application of technology to developing world agriculture is central to solving the problem of increasing food production.

These agricultural packages of the green revolution were, according to Conway and Barbier (1990), implemented in the most suitable agro-climatic regions and among those farmers who had the best means of maximizing the potential yield increases. Although heavily reliant on technology, chemicals and improved seed varieties, the resulting HEIA systems were productive both economically and in terms of alleviating food shortages (Pretty *et al.*, 1992; Conway, 1997). In addition, a wider process of 'commodification' took place, as agriculture shifted from a subsistence-orientated system to one determined solely by market forces (Dixon, 1990).

The major problem of adopting this approach, however, was that capital intensive HEIA was only suitable for those areas which met the ecological, marketing and transport criteria required for green revolution technology (Reijntjes *et al.*, 1992). Where these conditions were not met, none or very few of the packages were practical and consequently these remained unattractive to farmers. These unsuitable areas which include among others drylands, mountains and wetlands (Chambers, 1990), account for the majority of agricultural systems in sub-Saharan Africa and, according to Wolf (1986), a quarter of the world's population depend upon these areas for their livelihoods. Whereas the green revolution package produced a climate of economic growth in HEIA areas, the lack of investment in other areas resulted in over-utilization and environmental degradation as their population grew and, at the same time, farmers attempted to adjust to new market forces.

Over time, however, many of the successful HEIA areas also began to show signs of environmental degradation and economic losses for two main reasons. First, as a result of the repeated use of agro-chemicals and mechanization which were seldom designed to cope with a tropical environment, and secondly, the effects of only the partial adoption of green revolution packages as a result of the cost involved and the selective uptake of green revolution inputs by poorer households (Biot, 1989).

The general view is that this technocratic approach has done little to alleviate the widespread poverty and food shortages facing small farmers in many parts of the developing world (Binns, 1995). Researchers have also suggested that the green revolution itself has actually contributed to a widening of the gap between the rich and poor (Ladejinsky, 1969; Pearse, 1980) whilst also significantly increasing environmental degradation (Adams, 1990). More fundamentally, critics argue that the green revolution or scientific approach to development is outdated and inappropriate, on the grounds that:

1. most research and development has been carried out by professional western or western-trained researchers with an inherent cultural bias typified by inappropriate research methods (Chambers, 1983);
2. economic development has been pursued through technical fixes, which have in most cases resulted in inappropriate, unsustainable technology (Schumacher, 1973; Adams, 1990);
3. generally, a top-down approach to development has been adopted (Bartelmus, 1986; Adams, 1990). Governments and western scientists have agreed development policy, which has been communicated to development agents and the farmers carry out their instructions. Farmer consultation rarely occurs;
4. development initiatives have failed to look at the wider context beyond their economic goals. For example, the promotion of crop varieties which are economically or culturally unsuitable for the local population. Timberlake (1988) cites the case of a fast maturing sorghum species offered to Ethiopian farmers. This was rejected because the leaves and stalks were different to the natural variety and therefore it could not be used for roofing material;
5. there has been a lack of understanding of local socio-cultural systems on the part of researchers and, furthermore, a lack of participation by developers. Hirabayashi et al. (1980) point out that once development projects end, the target local population often revert to their original behavioural patterns;
6. there has been little recognition of failure, mainly because success has been measured by those who have a vested interest in carrying out development projects (Dixon, 1990);
7. development has, overall, concentrated on economic growth and increased agricultural production, with little consideration of the social consequences. As a result, the living conditions of the poor have not changed dramatically, even where development is considered an economic success.

According to Reid (1995):

The unsustainable impacts of current development trends mean that it is no longer possible to hold the view that the South's salvation lies in following the developmental path of the North (Reid, 1995, p.21).

With increasing recognition of the problems associated with the transfer of western style development to the developing world, researchers and development practitioners during the 1970s identified the need for a new approach. In particular, a recognition of the depletion of renewable resources in response to environmental and socio-economic pressure, prompted interest in the concept of sustainable agricultural development (Pearce, 1988). This included ideas of a more bottom-up approach to rural development, which sees the starting point of development as an equitable partnership between farmers, researchers and extensionists (Howes and Chambers, 1979). Subsequently, the concept of community participation also began to develop as, increasingly, existing sustainable agricultural practices were being recognized and attributed to local people's knowledge of farming practices and their environment (Chambers, 1983).

In 1980 Brokensha et al., published their work entitled *'Indigenous Knowledge Systems and Development'*. This emphasized the necessity for development planners to be aware of the vast knowledge and traditional skills held by the local people with whom they work. The result has been an increasing interest in the role of indigenous / local knowledge in rural development and the sustainable management of natural resources in the developing world, since the early 1970s.

Indigenous knowledge and development

McCall (1995) suggests that the documenting of indigenous technical knowledge (ITK) is not a recent phenomenon. During the 1930s, the colonial administration in British East Africa collected information on land use cropping practices and livestock rearing including methods of natural resource utilization (Roscoe, 1923; Trapnell and Clothier, 1937; Allan, 1949). According to McCorkle (1989) more systematic and detailed studies of indigenous knowledge systems began in the arena of anthropology during the 1950s, when researchers gave more attention to rural people's knowledge and its relationship to agriculture and ecology. In varying degrees, therefore, information on traditional, indigenous land use has always been available. It is, however, the attitudes towards IK which have changed significantly in recent decades.

Definitions and characteristics of indigenous knowledge

There is no standard definition of what constitutes IK. According to Warren (1996) 'indigenous knowledge' was initially used as a replacement for 'traditional knowledge' which implied simplicity and stagnancy, but now IK is open to a variety of misinterpretations. IK has become synonymous with local knowledge (AED, 1988), rural people's knowledge (Scoones and Thompson, 1994) and indigenous technical knowledge (IDS, 1979), although the latter originally focused on the role of people's knowledge and abilities in agricultural production, arguably obscuring the social and cultural dimensions of IK (Cornwall et al., 1994). According to Bell (1979) indigenous knowledge can be defined as:

> ...a mixture of knowledge created endogenously within the society and knowledge acquired from outside but then absorbed and integrated within the society (Bell, 1979, p.44).

In addition, it is said to be:

1. based on experience;
2. tested over centuries of use;
3. adapted to local culture and environment;
4. dynamic and changing (IIRR, 1996).

In the context of agricultural development, IK is characterized by practices which have evolved over time and are the result of a gradual learning process which emerges from observation, experimentation and the handing down of information from generation to generation (Chambers, 1983). Furthermore, IK has developed from a detailed understanding of local conditions and is modified in response to changing socio-economic, political and ecological conditions (Zwahlen, 1996). More fundamentally, the significance of IK lies in the ideological differences between it and western, scientific knowledge (Howes and Chambers, 1980; Warren, 1989; DeWalt, 1994) some of which are highlighted in Table 3.1. The two crucial underlying differences, which offer significant comparative advantages over the application of scientific knowledge, include first, the site and cultural specificity of IK and secondly, the ability of IK to adapt and evolve within this locality (Zwahlen, 1996). Consequently, it has been recognized that there is a need for development strategies to acknowledge and build upon these two key components of IK which, according to Fielding and Kirsop-Reed (1994), have sustained local communities for generations and particularly communities who have yet to meet any extension agents.

IK: the new panacea for development?

Historically, indigenous agricultural practices have been regarded as inefficient and an active hindrance to centralized, technically orientated development projects (Agrawal, 1995; Pretty and Shah, 1999). Since the 1970s, however, as the limitations of western style development have increasingly been recognized, rural development research has undergone a renaissance in terms of its interest in indigenous agriculture and it is now regarded by many to offer a greater potential for sustainable agricultural development (Chambers, 1983; Richards, 1985; Chambers *et al.*, 1989; Warren, 1991). Howes (1980), in one of the first analyses of the role of IK in development, identified the direct utility of IK as:

1. a shortcut to the compilation of an inventory of resources in a particular area, using local informants;
2. an environmental early warning system where indigenous observers can relay information to scientists;
3. the 'eyes and ears of science' whereby indigenous observers do much of the spade work in on-site scientific experiments (Howes, 1980).

Table 3.1 The contrasting ideologies of scientific and indigenous knowledge (adapted from DeWalt, 1994)

SCIENTIFIC KNOWLEDGE SYSTEMS	INDIGENOUS KNOWLEDGE SYSTEMS
Means used to study phenomena:	
Specialized, partial	General, holistic
Based on experiment	Based on observation
Resource utilization characteristics:	
Dependant on external resources	Dependant on local resources
High input	Low input
Land intensive	Land extensive
Labour saving	Labour demanding
Market risk	Environmental risk
Specialized adaptive strategies	Diverse adaptive strategies
Outputs:	
Low productivity for energy inputs	Low productivity for labour inputs
Cultural disjunctions	Culturally compatible
Profit goals	Subsistence goals
High potential for degradation	Low potential for degradation
Not sustainable	Sustainable with low population densities

This approach is rather limited in that the role of IK is portrayed as more or less an extension of scientific enquiry which can assist western orientated development projects. Howes (1980) does go on to suggest, however, that IK has the potential to play a more fundamental role, which includes the 'indigenous participation in the generation and exploitation of new techniques' (Howes, 1980, p. 344). This he claims would involve an interaction between farmers and scientists where both learn from each other and the outcome is an improvement in farming practices or technology. This idea has been developed significantly since the early 1980s and has evolved beyond the 'exploitative' approach. Increasingly, farmers have been recognized as acting upon the vast knowledge of their environment which they possess rather than being passive observers. The significance and utility of their IK rests more on their capacity to use and accumulate new knowledge in activities such as innovation and experimentation.

A great deal of research has focused on the contribution of the adaptive mechanisms of sustainable farming systems, namely the role of farmer innovation (Richards, 1985; Chambers *et al.*, 1989; Rhoades and Bebbington, 1995) and the transfer of information through indigenous communication channels (Winnarto, 1994; McCorkle and McClure, 1995; Mundy and Compton, 1995). The traditional view held by many scientists is one of farmers being adopters of technologies introduced from the outside, with any new innovations arising through accidental and unsystematic practices (Moris, 1991). The logic behind this assumption has been that the majority of farmers in the developing world practice subsistence agriculture and that any deviation from existing methods entails a high risk (Schultz, 1964).

Johnson (1972) was one of the first researchers to contest this traditional view of agriculture, maintaining that experimentation and risk are separate matters. Farmers can be cautious when faced with innovations but rarely do they refuse to try them out. Johnson (1972) goes on to argue that farmer experimentation in many instances, is an adaptive response to ecological variation although this is by no means a passive or uniform process. By responding to environmental changes each farmer contributes to the complexity of their knowledge system so that a critical degree of individuality exists among the knowledge base of farmers.

Richards (1985; 1986) describes several examples of experimental practices in rice farming among the Mende farmers in Sierra Leone. These farmers have areas of land reserved exclusively for experiments with rice varieties. In addition, rice is often planted in marginal areas such as the interzones between swamps and uplands, to test its tolerance to harsh conditions. Depending on the result, it can then be tested again either upslope or further into the swamp. Richards (1985) maintains that this is all part of a process of indigenous experimentation, which is designed through flexible research, to minimize the risks involved in environmental uncertainty. Furthermore, the study suggests that the complexity of this agricultural system and the environmental conditions under which it operates, do not lend themselves easily to standard modes of scientific investigation, hence the significance of farmers' IK.

Rhoades and Bebbington (1995), in their investigation of experiments undertaken by Andean farmers in the cultivation of potatoes, identified the existence of three types of experiments. Curiosity experiments are carried out when a farmer wishes to test an idea that comes to mind. They may or may not have a practical end and they usually occur when the farmer has time on his hands. Problem-solving experiments occur when farmers seek practical solutions to old and new problems and are more common in areas with diverse agriculture and poor extension services. Rhoades and Bebbington (1995) suggest that these experiments are ultimately superior to those possible as a result of the application of western science because farmers have the distinct advantage of having a greater opportunity to observe their agricultural environment. In addition, they are also in a much better position to prioritize those problems which affect them directly. The final class of experiments recognized are adaptation experiments which are carried out by farmers after they have acquired new knowledge or technology. These operate in three contexts:

1. when farmers are testing a new technology or idea in a known physical environment;

2. when farmers are testing an old technology or idea in a new environment;
3. when farmers test an unknown technology (Rhoades and Bebbington, 1995).

In addition to these types of experiment, Millar (1993) describes the existence of peer pressure experimentation among farmers in northern Ghana. This occurs when experimentation is initiated as a result of religious or cultural beliefs which see agriculture embedded in nature. Experimentation is equated with having respect for the land, hence it is believed that enthusiastic farmers who demonstrate originality are more likely to be rewarded with higher yields than those who repeat old experiments. Recently the 'cosmovision paradigm' has emerged from a series of studies (Haverkort and Hiemstra, 1999) which propose that farmers' knowledge and practices are based more upon a philosophy which is embedded in their own spiritual and holistic worldview, something which the TOT development paradigm has failed to appreciate.

A key issue raised by Rhoades and Bebbington (1995) in their classification of indigenous experimentation is that of the acquisition of knowledge. Whilst much of the IK debate has focused on the ability of farmers to generate and develop their own knowledge through experiments, a significant component of IK evolution and adaptation are the channels through which farmers acquire and communicate their knowledge. Whereas IK is created in the minds of individuals as a result of their perceptions of the environment, communication can be considered the process which enables people to create and disseminate new knowledge (Mundy and Compton, 1995). IK is generally communicated through events such as storytelling, village meetings and folk drama (Wang, 1982) in contrast to 'western' knowledge which is characterized by telecommunication systems, the mass media and government officials. Generally, indigenous communication systems are developed locally, are under local control, use low levels of technology and are also characterized by a lack of bureaucratic organization (Mundy and Compton, 1995).

McCorkle and McClure (1995) highlight a range of indigenous experimentation and innovation among Sahelian farmers in Niger, noting that farmers in remote areas were equally as innovative as those in more densely populated areas. Their research suggests that farmers place a much higher degree of credibility on the information they receive from fellow farmers than extension agents, especially when knowledge appears to have been well tried and subsequently adopted. Winarto (1994) reports a similar situation in Java, where farmers with rich agricultural experience gain respect from fellow farmers who willingly accept recommendations and information on farming practices. In McCorkle and McClure's (1995) study, western forms of communication such as the mass media and television were identified as playing some role in alerting people to new agricultural possibilities but interpersonal and group communication, and even direct observation, were found to be more significant in stimulating innovation. Within the study, farmers freely admitted that as sources of information agricultural extension agents with technical information were not relevant to their needs.

These studies demonstrate that IK is not shared equally throughout a society and, according to Swift (1979), differences can occur as a result of gender, age, experience and profession. Mundy and Compton (1995) suggest that sources of IK can be classified into five main groups:

1. indigenous experts who are skilled in areas such as crop production. These individuals are sought for advice on farming and other matters;
2. indigenous professionals who are experts with indigenous knowledge and skills not widely distributed in society. These may include healers, blacksmiths or irrigation system builders;
3. innovators who deliberately experiment and develop new knowledge;
4. intermediaries who include those responsible for informing members of society about new ideas or developments;
5. recipient disseminators who are individuals who receive new information then react to it before passing the knowledge on.

These indigenous communication sources are highly significant in the adaptation and evolutionary process of IKS. Unlike most exogenous communication channels which are met with scepticism on the part of farmers, they are regarded as highly credible sources of information which are significant facilitators of innovation and change. In addition, they are the means by which IK is preserved and passed down from one generation to the next (Mundy and Compton, 1995). They are, therefore, of fundamental importance in any form of rural development which relies on the dissemination and diffusion of information among local communities. As studies have shown (McCorkle and McClure, 1995), the communication of knowledge and ideas through exogenous channels alone may have limited success.

What this brief discussion of farmer innovation and experimentation highlights is that IKS are both flexible and dynamic. There is overwhelming evidence from around the world to suggest that in many cases, small scale indigenous agriculture is characterized by practices which have been and continue to be relatively sustainable (Richards, 1985; Chambers *et al.*, 1989; McCall, 1995, Haverkort and Hiemstra, 1999). Furthermore these farmers (and their IKS) have the capacity to acquire and generate IK as part of an adaptive process which can occur in response to environmental, socio-economic and cultural changes. In particular, farmer innovation is primarily regarded as a means of problem solving resulting in environmental adaptation, although, significantly, it also appears to be part of a continuous process of improvement.

In response to the question of how development can build upon and incorporate these indigenous mechanisms and structures, there is widespread agreement that farmer participation in agricultural research constitutes the way forward (Farrington and Martin, 1988). Chambers *et al.* (1989) recommend that development should take a 'farmer first' approach in which rural people's knowledge and problems take priority and the farmer becomes the principal researcher in the development process. Within this development paradigm the role of the development agent is seen as a facilitator who brings together IK, strengthens the indigenous communication network and promotes the exchange of information, thereby strengthening the capacity of farmers to produce solutions to their own problems. Röling (1994) argues that in this way facilitation is the key element to sustainable agriculture. Rajasekaran (1993) meanwhile argues that this approach can only be operationalized through the documenting and recording of IK, establishing IK resource centres and the provision of relevant training for development practitioners. Critics such as Agrawal (1995), however, argue that such an approach would be inappropriate to IK which is essentially dynamic and both environment and culture specific.

The initial implications for development are that the sustainable utilization of natural resources appears to be attainable if research and extension incorporates, facilitates and is sensitive to the many aspects of IK. Reijntjes *et al.* (1992) make the point, however, that if indigenous farming systems function as well as supporters of IK suggest, then it is a mystery as to why extensionists and development workers are required to intervene in the first place.

The limitations of IK

Several authors have pointed out that traditional methods of natural resource management based on IK may have been sustainable in the past (e.g. slash and burn cultivation) but the rapid environmental and socio-economic changes which have occurred in recent years, have exceeded the capacity of farmers to adapt and cope (Farrington and Martin, 1988; McCorkle, 1989; Wood, 1991; World Bank, 1998). In particular, Farrington and Martin (1988) suggest that IK may have problems adjusting to cyclical variations in climate, such as those experienced in the Sahel.

These views are part of an overall recognition that a dependence on IK alone cannot solve current development problems and that IK has several inherent limitations. These include:

1. IK is a community resource which tends to be distributed unevenly among its members. The possession of knowledge may relate to economic or cultural activities carried out by specific individuals (Reijntjes *et al.*, 1992). For example women may possess greater knowledge than men on the medicinal use of herbs (Shiva, 1988). Furthermore, those members who possess more IK may not necessarily wish to pass on their knowledge to either their colleagues or to external researchers. There is, therefore, some doubt as to whether the elicitation of IK can be anything more than fragmentary (Fielding and Kirsop-Reed, 1994);
2. it has been argued that because IKS are rich in contextual detail, they are non-transferable, being of little use outside their own population;
3. there is a general criticism that farmer experimentation is undirected and uncertain in comparison to 'formal' science. Juma (1987) argues, however, that farmer experimentation cannot be viewed from a reductionist viewpoint in the same way as scientific experimentation, principally because farmer experimentation is embedded in a specific cultural context. Richards (1989)

takes this further, suggesting that agriculture needs to be viewed more as a performance at a specific point in time under specific circumstances, rather than the result of a closely controlled, calculated set of objectives.

As a result of these criticisms, most researchers now agree that IK alone cannot be relied upon to solve development problems. At the same time, there has also been an acknowledgement that the application of the scientific paradigm is not entirely redundant in the arena of rural development. Zwahlen (1996) suggests that the dichotomy between IK (= sustainable) and scientific knowledge (= unsustainable) is too simplistic an approach. It is the interaction of both, resulting in the application of knowledge which reflects a diversity of cultures, which may provide the key to success.

Towards a global knowledge system

In Brokensha *et al.* (1980) several authors recognized that IK and scientific knowledge could be complementary, especially in farming research where the strength of ITK lay in its empirical nature, which was seen as an extension of the scientific process (Howes, 1980). In his ethno-ecological research on community understanding of the effects of grasshoppers in Nigeria, Richards (1980) clearly demonstrated the advantages of integrating both IK and scientific knowledge, in terms of a combination of the empirical advantages of IK and the inductive logic of scientific enquiry. The relationship between the two knowledge systems was classified under four main headings:

1. Where the research team findings transcended local knowledge.
   These results tended to depend on precise, quantitative data, experimental control and biochemical analysis.
2. Where community knowledge hinted in general and imprecise ways at results obtained by researchers.
   For example, where population fluctuations of grasshoppers were found to be related to climatic changes. In this case, local knowledge was found to be the equivalent of the initial hypotheses with which the scientific researchers started.
3. Where local knowledge was equivalent to research team findings.
   For example, where fungus and fly larva were found to influence the mortality among grasshoppers as the rains began. Here the farmers described egg laying behaviour of the grasshopper and some suggested methods of eradication based on this information.
4. Where local knowledge was additional to research team findings.
   This occurred where communities had detailed knowledge unavailable to the researchers, including information on the history and geographic location of grasshopper outbreaks (Richards, 1980).

In both (1) and (2) the overlap exists as a result of observation being the crucial, common factor, whereas (1) demonstrates the fact that farmers do not have access to scientific equipment. Alternatively, the researchers were equally deficient in their knowledge in (4) because they were not living as part of the farming community. Clearly in this research context both types of knowledge appear to have their strengths and weaknesses, stressing the need to maintain a diversity of approaches.

The more recent trend in IK research, as suggested earlier, has been to place more emphasis on the dichotomy between scientific knowledge and IK, maintaining that IK offers more than simply contextual data for science (Howes and Chambers, 1980). Consequently, the utility and integration of both has become much more significant. DeWalt (1994) reviews several case studies which highlight the limits of both scientific knowledge and IK, concluding that they should be viewed as complementary sources of wisdom. The key issue is providing both systems with more opportunities in which they can inform and stimulate each other. DeWalt (1994) also suggests that resource utilization strategies should not rule out external inputs such as fertilizers and machinery. Instead, the emphasis should be on the cultural compatibility of proposed scientific solutions, which can be achieved through the interaction of indigenous and external actors.

This process of interaction is described by Ryden (1991) who proposes a 'loop' model for development which incorporates the expertise inherent in both IK and western approaches. This model applies to natural resource management systems which are, for the most part, unsustainable and showing signs of environmental degradation. The three stages of the model represent steps towards regaining some degree of sustainability through the interaction of local communities and external researchers:

1. external researchers attempt to understand the logic and ecological stability of traditional resource management systems and their associated IK;
2. using local and external expertise, the reasons why indigenous management is no longer adequate are identified;
3. local and external expertise interact to develop innovations which can solve the specific resource management problems (Ryden, 1991).

Ryden (1991) suggests that following a mutual understanding of the problems involved, the characteristics of indigenous management methods can be related to scientific concepts. Innovations which incorporate aspects from both knowledge systems can then be tested in the field with the local communities.

Similarly, Rajasekaran *et al.* (1991) propose an integrated natural resource management system model were the aim is to increase food production with the least amount of resource deterioration, by incorporating both indigenous technologies and economically feasible 'outside' technologies. The starting point is the recording of IK which identifies those existing practices which are well adapted to the local conditions. This is widely regarded as a cheaper, faster and more fruitful method of knowing the spatial location of resources in an area (Atteh, 1989). The economic feasibility of new practices and whether these can be incorporated or modified by IK is then evaluated and implemented. Throughout the implementation of these new strategies the aim is to incorporate IK into their running, in response to any changes in local conditions. Rajasekaran *et al.* (1991) maintain that this approach can increase agricultural productivity while also conserving the resource base and strengthening the role of IK.

In both examples (Ryden, 1991, Rajasekaran *et al.*, 1991) the primary role of science is to comprehend the logic of IK and the degree to which it is adapted to environmental conditions. Through the application of its own methods, science can ascertain the extent to which indigenous perceptions are an 'accurate' representation of environmental conditions, thereby identifying the root of any maladjusted or ineffective management techniques.

McCorkle (1989, p. 4) considers a 'knowledge of local knowledge', in particular knowledge of the rich, often site specific IK, to be the first stage in any development initiative. In the broader context, several authors argue the need for IK to be extensively documented so that it becomes a resource to developers in much the same way as science (Zwahlen, 1996; Rochleau, 1987). Knowledge and information from any source can then be chosen and applied in combination, wherever necessary. Warren (1996) sees this more as a shift from specific knowledge systems to a global knowledge system which becomes a resource for the global community.

This 'global' approach to knowledge has become more popular, as increasingly the need to describe, investigate and document IK has become a fundamental part of development projects. In 1998, the World Bank published 'Indigenous Knowledge for Development: a framework for action' which aimed to facilitate the incorporation of IK in development initiatives through a process of participation and, in particular, learning on the part of developers. The four key areas of this framework are summarized in Table 3.2.

Table 3.2 A framework for incorporating IK in development (adapted from World Bank, 1998)

1. Disseminating information:
   - Developing a database of IK practices, lessons learned, sources, partners, etc.
   - Identifying and testing instruments for capture and dissemination of IK.
   - Publishing selected cases in print and electronic format.

2. Facilitating exchange of IK among developing communities:
   - Helping build local capacity to share IK, especially among local IK centres.
   - Identifying appropriate methods of capturing, disseminating IK among communities.
   - Facilitating a global network to exchange IK.

3. Applying indigenous knowledge in the development process:
   - Raising awareness of the importance of IK among development partners.
   - Helping countries to prepare national policies in support of indigenous practices.
   - Integrating indigenous practices in programs/projects supported by partners.

4. Building partnerships:
   - Learning from local communities and NGOs.
   - Leveraging limited resources of partners to obtain greater development impact.
   - Addressing the intellectual property rights issue of indigenous knowledge.

It was proposed that the use of such a framework offers the potential for increased awareness of IK and a more global knowledge system. Its critics suggest that such a focus on the practicalities of incorporating IK with western knowledge inevitably result in only the empowerment of those aspects of IK which *can* be readily understood from the perspective of western knowledge (Haverkort et al., 1999). Recent IK research (Millar, 1993; Kieft, 1999) has instead suggested that the key to understanding the whole of IK lies in local people's cosmovisions, i.e. the way in which they perceive the world, whether this be natural, cultural or spiritual. In addition, 'endogenous development' has emerged which builds upon the principles of 'farmer first' but within the cosmovision paradigm. This is seen as development from within, characterized by the strengthening of indigenous institutions which facilitate the participatory process among farmers. Gonese (1999) suggests that the role of 'outside' development projects should, therefore, be political rather than technical, in a way which, as suggested by Thrupp (1989b), empowers local communities by legitimizing their IK.

The dichotomy of scientific and indigenous knowledge

In examining the relationship between IK, external knowledge and development, it is useful to view this in the context of a wider academic debate. While there is now general agreement amongst researchers, developers and academics that both IK and science should be an integral part of rural development, there appears to be an ongoing epistemological debate which questions the validity of the whole IK and scientific knowledge dichotomy. In his critique entitled 'Dismantling the divide between indigenous and scientific knowledge' Agrawal (1995) argues that although researchers constantly claim that IK offers a new approach to development, it does in fact differ very little from the technical, scientific approach. The generally accepted basis for this divide has been on:

1. substantive grounds – because of differences in the subject matter and characteristics of indigenous and western knowledge;
2. methodological grounds – they employ different methods to investigate reality;
3. contextual grounds – because traditional / indigenous knowledge is more rooted in its context.

The substantive argument maintains that IK is concerned with activities surrounding the livelihoods of people rather than the abstract ideas and philosophies of science, which appear to be divorced from the daily lives of people. The growing body of IK literature and case studies, however, suggests that IK is itself characterized by insights, ideas and innovations. Agrawal (1995) also argues that it is inaccurate to suggest that science and western knowledge do not touch the everyday lives of people in the developing world. In this respect there is little difference between science and IK.

Secondly, it is argued that whilst science is systematic, objective and analytical, IK is closed, subjective and little more than common sense. In response to this,

Agrawal (1995) claims that philosophers of science recognize that there is still no methodology to distinguish science from non-science (Kulka, 1977). The separation on methodological grounds is, therefore, erroneous. To say that IK is also closed would also be an error. Thrupp (1989a) observed a range of attitudes displayed by local populations towards new knowledge, including pride in traditional methods, rejection and acceptance of new knowledge and embarrassment of old methods.

The contextual argument separating IK and science claims that IK exists in harmony with the lives of those who generate it, whereas western science thrives on abstract formulation, beyond the scope of people's lives. Here, Agrawal (1995) argues that in the past development policies have failed because they tended to ignore the socio-cultural contexts in which they were implemented. If this is true, it is logical to assume that western science is itself too deeply rooted in its own context. In this respect, there is little difference between IK and western scientific knowledge.

Agrawal (1995) suggests that the recent interest in IK by the scientific community is more or less doomed to failure primarily because researchers are still clinging to the dichotomy of western knowledge *vs.* IK when in fact this is not true. Secondly, he claims that this is being reinforced by the fact that IK is being collected and conserved outside the area to which it belongs. In most cases research into IK is ending up on the shelves of resource centres in the developed world, which remain largely inaccessible to anyone but the scientific community (Heyd, 1995). One of the most important features of IK is that it is both site specific and dynamic but the current approach to its study and conservation succeeds in isolating it. In Agrawal's words:

> ...ex situ preservation of IK is likely to fail, succeeding only in creating a mausoleum for knowledge. Even if it is successful in unearthing useful information, it is likely to benefit the richer, more powerful constituencies – those who possess access to international centres of knowledge preservation (Agrawal, 1995, p. 432).

In assessing the way forward for IK, Agrawal (1995) argues that those whose principal interests lie in conserving the diversity of IK, should channel their energies towards reorientating and reversing state policies and market forces so that the holders of IK can determine their own future. Holders should have full control over their lands and resources and if this is achieved IK can continue to be an integral component of natural resource management.

Indigenous wetland management

Having reviewed the debate surrounding the relationship between IK, western knowledge and their role in development, the utility of IK in the context of wetland management is now addressed. In Chapter Two, wetland development was discussed from the point of view of natural resource degradation, with the emphasis on the increasing threats to wetlands and the need for effective, sustainable management. The chapter also drew attention to the lack of wetland research or management policies which go beyond conservation or development for a single utility and address the issues of human interaction and utilization of the world's wetland resources. This chapter, however, has argued the need to incorporate an appreciation of IK into research and development initiatives, so that indigenous management practices and knowledge can be empowered, stimulated and developed to meet the demands of a changing environment, whilst facilitating some degree of sustainability.

Until relatively recently, examples of research which have addressed the role of IK in wetland management remain scarce and under-reported. Despite this fact, the examples highlighted in the following sections clearly demonstrate that a consideration of IK should be fundamental to any wetland management initiative.

The role of IK in wetland management

The history of wetland development in much of the developing world mirrors that of agricultural exploitation. Evidence from around the world suggests that the IK of wetlands has been developed over centuries through a process of trial and error, resulting in a range of management techniques which are specifically adapted to the local environment (Nkwi and Toornstra, 1989). The stability of these management systems has been disrupted as a result of external pressures and development initiatives which have in many cases led to the over-exploitation of wetlands and the abandonment of indigenous practices which were previously well adapted (Diegues, 1989). A clear disregard for indigenous management techniques has had a dramatic effect on many of Uganda's valley swamps (Tukahirwa, 1989; Denny and Turyatunga, 1992).

Studies which highlight the effects of development and the TOT approach on wetlands and their communities are ubiquitous (Hollis, 1990; Adams, 1993; Turner, 1994). Less common are those which clearly demonstrate the role of IK in wetland management. As Gopal (1989) points out, the traditional view held by western developers has been that indigenous attempts at wetland management are primitive and static, whereas in reality, wetland management can only be successful if it includes local people, their knowledge and their capacity to adapt to changes.

Drinkwater (1994) highlights the adaptive nature of *dambo* cultivation in Zambia, where the spatial and temporal distribution of practices reflect the variation in agronomic problems posed by *dambo* soils. The rationale behind many of these practices, in particular the selective burning of wetland soils, has yet to be understood scientifically although such practices appear to be rooted in farmers' knowledge base.

Bell and Hotchkiss (1989) describe how, during wet years, *dambo* farmers construct drainage channels and farm raised beds to attain suitable soil moisture conditions. This is adapted during drought conditions to a system of irrigation using water obtained from wells. These practices represent a flexible extension of dryland agriculture, which otherwise could not cope with the demand for food during the 'hungry season'.

Fairhead and Leach (1994) describe indigenous swamp management practices in Guinea, where in particular, farmers maintain water levels by encouraging the growth of specific trees and weeds. Sietchiping (1998) also reports on how farmers in Cameroon have recently begun to experiment with wetland agriculture in response to various socio-economic pressures. These farmers have developed a system of wetland agriculture which includes a range of soil and water conservation measures, a variety of crops and crop rotation practices.

Although not referring specifically to the role of IK in wetland management, much research has been carried out on the floodwater farming systems of West Africa (Kimmage and Adams, 1992; Adams, 1993a, Hollis et al., 1993). These farming systems are characterized by an extension and adaptation of dryland agricultural practices to the marginal areas of wetlands and have been demonstrated to be sustainable over a long period of time. Farmers have acquired an intimate knowledge of their environmental conditions, in particular the seasonal hydrological changes and the implications of these on soil fertility and crop growth. As a result, adaptations to the changing hydrological regime are commonplace as well as a variety of risk minimization strategies.

The ability of IK to adapt to changing circumstances is demonstrated by Zimmerer (1991) who presents an account of the rise of indigenous wetland management among Peruvian farmers in the Andes, which has come about as a result of demographic and socio-economic pressures on dryland farming areas. Although marshy areas were previously utilized only for grazing, farmers were able to adapt to this new farming environment within a matter of years. Zimmerer (1991) argues that initially farmers' knowledge of dryland agriculture was applied to wetland areas and subsequently modified in an active process of innovation and adaptation. In addition, environmental monitoring of the extreme spatial and temporal variation in soil moisture conditions played a critical role in determining factors such as the allocation of labour or the location of drainage channels. Within several years of cultivation wetland agriculture had become commercialized to the extent that other agricultural activities became marginalized. Zimmerer (1991) considers this a prime example of successful development originating from the utilization and adaptation of stocks of IK. Furthermore, the study demonstrates the ability and versatility of IK to cope with significant environmental or socio-economic changes, contrary to the view that adaptation may, at the very least, be problematic on already established farming systems.

The hazards of development which fail to recognize the critical role of IK in wetland management are highlighted by Dries (1989) who presents a review of wetland rice management systems in Sierra Leone. Dries (1989) reports that these indigenous wetland management systems have been in operation since the end of the 19th century. Since then, as wetland farming innovations have spread throughout each community, the basic practice of rice cultivation in these upland valley swamps has evolved to include among other things fallow periods and a dry season crop which can be grown on mounds in each wetland.

In response to increasing demands for rice, however, the government sought a state of self sufficiency which it pursued through the intensification of agriculture and the application of green revolution packages from 1986. Paying little attention to existing indigenous wetland management systems, the government tried to introduce a completely new system which involved the levelling and restructuring of wetlands, along with the provision of equipment to farmers. Dries (1989) suggests that this new system failed for three reasons. First, the new cultivation techniques were too labour intensive for local conditions because the ratio of farmers to cultivated land was low. Secondly, the technical packages provided were not sustainable because of their expense, and thirdly, the drainage system was not adapted to local hydrological conditions, hence the wetlands dried out and most were abandoned by farmers. The ultimate responsibility for failure however, clearly rested with the government who chose to ignore the contribution of IK in a well adjusted wetland management system.

Empowering indigenous wetland knowledge

In a similar way to most agricultural resources, the obvious way forward for wetland development is the adoption of a 'farmer first' or 'cosmovision' approach which is culturally and spiritually sensitive, and which recognizes and values the range of indigenous wetland knowledge held by local communities. Mermet (1989) makes the critical point, however, that wetland management is a process in which many stakeholders with different goals are participating and that wetland management should not be dominated by a single approach. Although much of the research on the role of IK in agricultural development has focused on IK as a community resource, in practice the focus has been on utilizing IK at the level of individual farmers. In contrast, wetland management, if it is to be sustainable, requires an understanding of the range of individual goals. This means not only those goals of local communities whose livelihoods depend upon wetlands but also those who consider themselves to be wetland managers, in particular, government planners or NGOs. As Mermet (1989) concludes, empowering people and their knowledge in the wetland management process requires the vision to involve all the actors at all the different management levels. This requires adopting a range of approaches which are sensitive to the knowledge system of each.

Similarly, Oakley and Driyver (1989) suggest that local people's knowledge of wetlands should be understood but not allowed to become a dogma. External knowledge, which often recognizes the value of wetlands in different ways to indigenous communities (e.g. biodiversity) should be equally entertained, leading to the participation of all the wetland stakeholders in the management process. Examples of wetland management initiatives which have adopted this participatory approach, however, are relatively scarce.

Noble (1996) describes how increasing pressure on farmland has led to the over-exploitation of traditionally cultivated *dambos* in Malawi. Through a participatory process involving farmers, the government and an NGO, new wetland management systems were developed which incorporated fish ponds, rice and vegetable gardens, depending upon the range of local conditions. Furthermore, initial ideas and stimulation were taken further and modified to the point where farmers engaged in their own process of experimentation through which their regenerated wetland management system is constantly evolving.

More recently, the role of local people in wetland management has been addressed at the 2nd International Conference on Wetlands and Development, held in 1998 in Senegal. One of the five key themes of this conference was the participatory management of wetlands and the identification of the opportunities for participation between governments, NGOs and local communities (Gawler, 2002). In particular, it was acknowledged that the empowerment of IK, although it does not guarantee sustainability, should be used as a complementary source of knowledge to that of the scientific community and other wetland stakeholders. The current emphasis should be on the bilateral exchange of information and the participation between these different groups (Wetlands International, 1998).

In addition, the Ramsar Bureau in their 1999 conference entitled 'People and Wetlands: The Vital Link' recognized that indigenous peoples and local communities are already involved in managing and using wetlands in a sustainable manner. Debate within this conference also centred on the need to identify the range of wetland stakeholders and include them in the decision-making and management process (Ramsar, 1999). As a result, Resolution VII.8 established guidelines for strengthening local communities' and indigenous people's participation in the management of wetlands. It remains to be seen whether this recognition will ultimately result in the inclusion of indigenous people and their knowledge in the wetland management and planning process, or whether this approach merely seeks their co-operation in order to legitimize existing conservation focused strategies.

Conclusions

The application of the scientific paradigm to development has meant that the knowledge held by local communities, which has evolved in close contact with their environment, has in most cases been undervalued and ignored by developers and natural resource managers in the past. Research over the last twenty years suggests, however, that this indigenous knowledge often has the capacity to adapt and cope with environmental changes and in this respect it represents a form of sustainable management. The speed of change in many parts of the developing world has occurred at such a rate that IK cannot cope and the result in many cases has been environmental degradation and widespread poverty for rural communities.

In response, researchers have suggested that development strategies should include a 'farmer first' approach which places greater emphasis on the role of local communities and their IK in their own development. The aim is to empower IK but it is also important that scientific knowledge is not ignored. The development process should be participatory but this should involve the participation of all the wetland stakeholders so that different knowledge can be exchanged, developed and adapted to meet individual needs. Ideologically this represents a shift from several knowledge systems (e.g. science and IK) towards the application of a global knowledge paradigm in the development process.

While development has been pre-occupied with technocentric, western style development, wetland managers have similarly been pre-occupied with large wetlands, supporting rich biodiversities with the emphasis on conservation. Although local communities have been recognized as users of wetlands for the past 15 years, only recently have they been regarded as managers who perhaps possess the knowledge to manage these resources in a sustainable manner. Building upon this recognition, there is at present a significant opportunity to apply the lessons learned from past rural development initiatives, IK research, wetland development experiences and participatory research methodology to the development of sustainable wetland management strategies, especially in those areas where wetlands represent an essential natural resource to local communities.

As discussed in Chapter Two, wetlands are critical to the survival of many communities throughout the developing world and in many parts of Africa research has drawn attention to the need for sustainable management. In contrast, little is known of Ethiopia's wetland resources, primarily because the country has remained relatively closed to outsiders in the past but also because it has never been a signatory to the Ramsar Convention on Wetlands of International Importance.

Ethiopia's wetlands, which undoubtedly have at least the same significance for local communities as the better documented wetlands in other parts of Africa are possibly under a greater threat of exploitation because they are neither recognized as important by their own government or by international agencies. On the other hand, this lack of recognition may have been significant in preventing their externally generated development, presenting a unique opportunity to investigate the indigenous utilization of wetlands.

Chapter 4

Wetland Resources in Illubabor

Introduction

In this chapter the study area of Illubabor in South-western Ethiopia and the nature of its wetland resources are introduced. Despite receiving little national or international attention, the relatively small valley bottom wetlands located in the south-western highlands are of critical importance to many rural communities in terms of the range of natural and socio-economic benefits they provide. In view of this importance, there has been some concern over the possible effects of an intensification in wetland use, particularly towards their drainage for agriculture. Where this intensification in wetland use seems to have occurred, reports suggest that the benefits have been short-lived and mostly unsustainable.

These wetlands are, however, part of a wider environmental and socio-economic context in which an array of influences determine the extent to which they are utilized and whether their use remains sustainable or not.

Ethiopia: the physical setting

Ethiopia is situated in the north-eastern part of Africa, between the latitudes 3° and 18° N and longitudes 33° and 48° E. It is bordered to the west by Sudan, to the south by Kenya, to the south and south-east by Somalia and to the north and east by Eritrea and Djibouti (Figure 4.1). The total area of the country is approximately 1.27 million km^2, with an estimated population of around 59.8 million for 1998 (CSA, 1997). The country can be broadly divided into two main areas: the highlands and the lowlands, above and below an altitude of 1500 m. The highlands cover approximately 44 per cent of the country's area and are the most significant area economically, supporting over 85 per cent of the country's population and generating more than 90 per cent of its agricultural output (Tagoe, 1983).

The climate of Ethiopia is a result of this topographical variation and its position in Africa. Where the terrain is above 2400 m average daily temperatures range from near freezing to 16°C, with March, April and May constituting the warmest months. In the lower areas of the plateau between 1500 and 2400 m the daily temperature can range between 16°C and 30°C. Those areas below 1500 m, such as the Danakil depression in the north-east, the Ogaden desert in the east and the periphery bordering Sudan and Kenya exhibit greater fluctuations with temperatures often in excess of 40°C. Rainfall is also variable throughout the

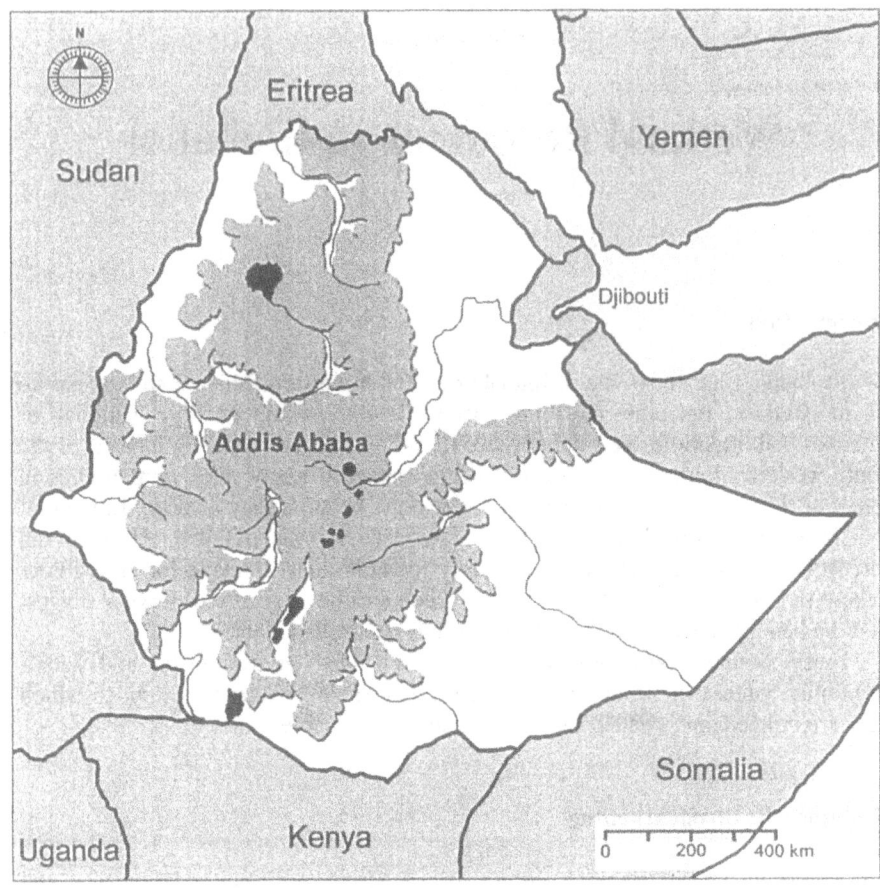

Figure 4.1 The location of Ethiopia and the Ethiopian highlands

country as a result of differences in elevation and the seasonal changes in atmospheric pressure which control the prevailing winds (Figure 4.2). Between October and January the Inter Tropical Convergence Zone's (ITCZ) southern position brings an extensive dry season to much of Ethiopia. The ITCZ gradually moves northwards between March and June, facilitating the movement of monsoonal air from the south-west. The primary rainy season, known locally as *kremt*, then begins in mid-June and lasts through to September. This main rainy season is, however, usually preceded in April and May by converging north-east and south-east winds, which produce a brief period of rains in northern, central and eastern areas known as the *belg* (McCann, 1995).

As a result of the wide range of climatic conditions and the altitudinal variation within the country Ethiopia benefits from a diversity of natural resources which sustain a range of rural economies including animal rearing, mixed farming systems and pastoralism (Wood and Stahl, 1989). Subsistence agriculture remains the most ubiquitous livelihood for Ethiopia's population although soil erosion

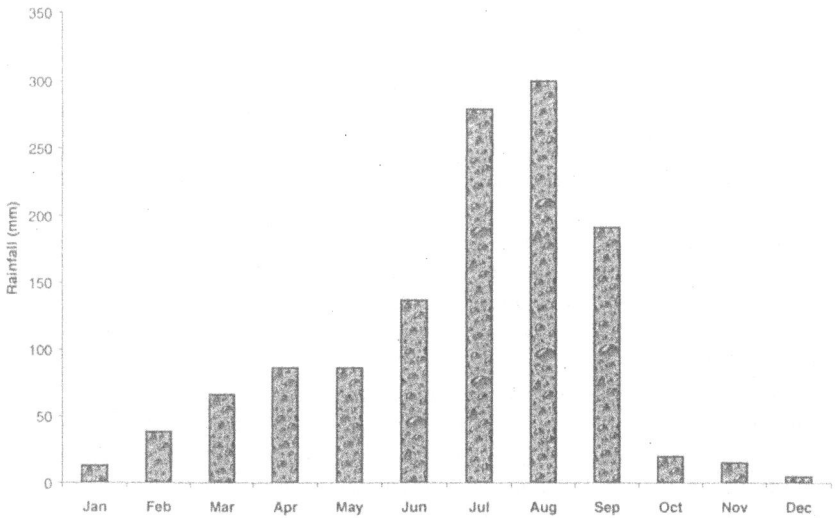

Figure 4.2   The seasonal pattern of rainfall in Ethiopia

poses major problems for farmers as a result of the steepness of slopes under cultivation, the high rainfall intensity and the farming practices themselves (Hurni, 1988; Wood, 1990). The problem of soil erosion has been intensified in many areas, especially in the north of the country, where the land has suffered from deforestation and drought is a regular occurrence (Alemneh Dejene, 1990). Nonetheless, many of the soils found throughout Ethiopia have a high fertility, offering significant agricultural potential.

Although economic and population pressure have led to deforestation as demand for timber, pasture and farmland has increased, highland Ethiopia still retains some areas of coniferous or broadleaved forest. According to the FAO (1995) natural forest accounts for approximately three per cent of Ethiopia's total land area, although this represents only a fraction of the estimated forest landcover at the end of the 19th century (approximately 40 per cent). Nonetheless, these remaining areas support a range of activities including coffee cultivation and foraging, in addition to their direct exploitation for timber. In addition, they constitute some of Ethiopia's most important areas for genetic and wildlife resources, sustaining many endemic species of flora and fauna.

Ethiopia's water resources include a number of Africa's major rivers. Wood and Stahl (1989) suggested that the total runoff exceeds 100 billion cubic metres, half of which comes through the Blue Nile system, 14 per cent through the Omo river in the south and 12 per cent through the Baro Akobo system in the south-west. Given the general westward slope of the highlands, most of the large rivers in Ethiopia are tributaries of the Nile basin. The country's largest river, the Blue Nile, accounts for nearly two thirds of the total Nile river flow downstream in Sudan and Egypt.

In the highland areas most rivers are deeply incised, making them significant resources in the production of hydro-electric power. In contrast, the lowland rivers offer greater potential for irrigation schemes. Wood and Stahl (1989) estimated the total potentially irrigable area as approximately 2.24 million ha.

Apart from the river systems, Ethiopia's wetland resources include a number of lakes located throughout the country of which Lake Tana in the north constitutes the largest and forms the source of the Blue Nile. The rift valley lakes to the south of the country are, in contrast, of tectonic or volcanic origin and are characterized by closed drainage basins (Tudorancea et al., 1999). The littoral zones of many of these water bodies are characterized by swamp vegetation including *Cyperus* sp., *Typha augustifolia* and *Phragmites* sp. as are a variety of seasonally and permanently flooded wetland areas which occur in most of the country's agro-climatic zones.

In particular Gambella zone, situated to the west of the country, is characterized by extensive seasonal swamps which are contiguous with those of the Sudd region of Sudan. This area along with the wetland areas around Lake Langano and Lake Awasa have recently been designated national parks by the Ethiopian government. Despite this wealth of water resources there has been limited exploitation of these areas for fish production and consumption.

Illubabor zone

Illubabor zone is situated to the west of the country (Figure 4.3) and remains one of the most fertile and least exploited regions in Ethiopia. It has a total area of approximately 16,555 km^2 and lies between longitudes 33°47' and 36°52' east and latitudes 7°05' and 8°45' north (CSA, 1997). It is bordered to the north by East and West Wollega zones, to the south by Kefa zone, to the east by Jimma zone and to the west by Gambella zone. Following the overthrow of the communist government in 1991, the Republic of Ethiopia was, in 1993, divided into 14 regional states, broadly reflecting the spatial distribution of ethnic groups throughout the country. Illubabor is one of 12 zones within the regional state of Oromia and is itself composed of 12 *weredas* (districts) (Figure 4.4). These *weredas* represent the smallest administrative unit in which government salaried officials are involved, although one further government sub-division exists: that of the *kebele* (formerly the peasants association) whose residents meet to discuss community issues and agricultural practices.

Relatively little information is available on the geographical characteristics of Illubabor zone beyond that which can be elicited from the various maps available, although several studies have been carried out in and around the area. These include studies by Wood (1977), Solomon Abate (1994) and more recently research carried out by the Ethiopian Wetlands Research Programme (EWRP) (Tegegne Sishaw, 1998; Solomon Tekalegn, 1998; Asmamaw Legasse, 1998). In addition, a variety of unpublished government reports from the Ethiopian Ministry of Agriculture (MoA) contain some information on the area's environmental

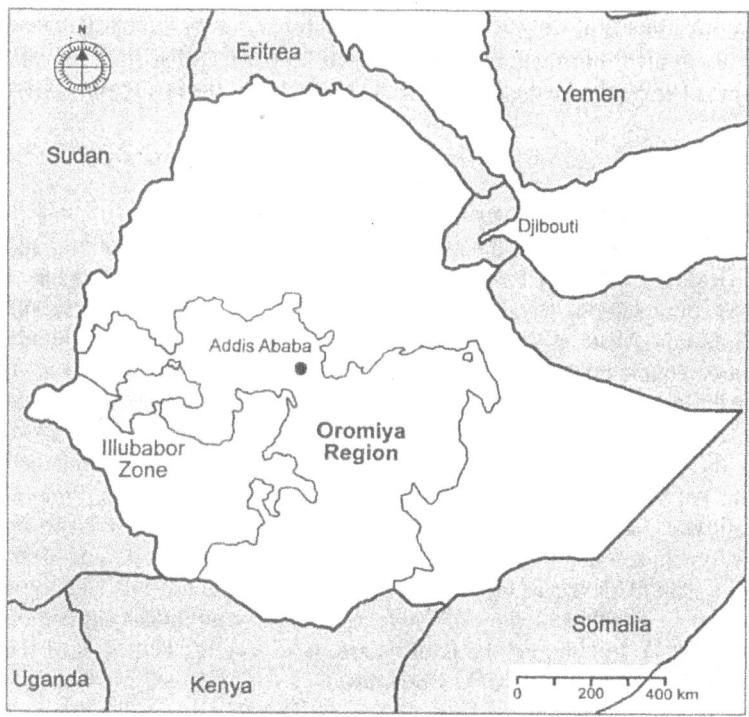

Figure 4.3 The location of Illubabor zone within Ethiopia

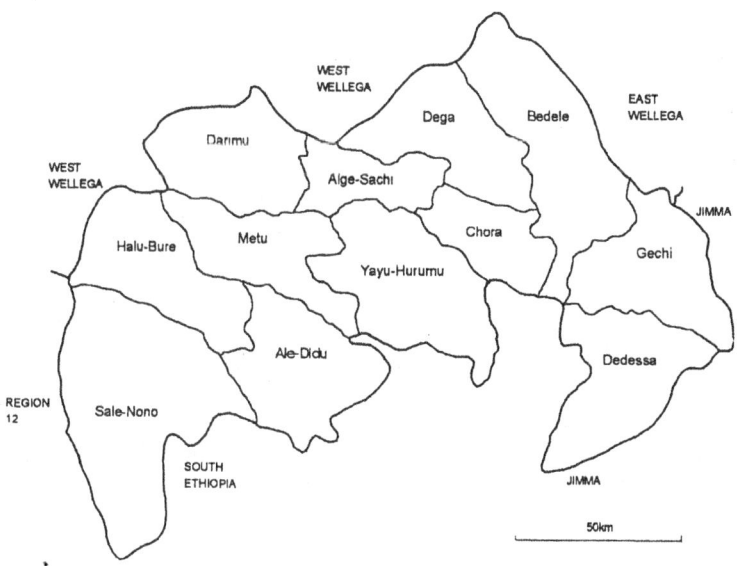

Figure 4.4 Administrative divisions (*weredas*) within Illubabor zone

resources (although this is of varying quality and reliability due to inconsistent data collection). This limited information is compounded by the fact that Illubabor has been the subject of several boundary and district changes over the last thirty years.

Physical characteristics

The geology of Illubabor is dominated by the Precambrian basement complex to the north and west of the zone and the more recent basalt lava flows in central and eastern areas (EMA, 1987). The basement complex is typically characterized by a sharply crested, mountainous terrain whereas that of the basalt is more hilly and undulating (Solomon Abate, 1994). The topography is generally that of a highly incised plateaux characterized by dendritic drainage systems, which ranges in altitude from 800m at the bottom of steep sided river valleys to 2500m at the mountainous southern boundary of the zone.

The soils throughout the zone consist mainly of nitosols on the basalt and lixisols on the basement complex. Soils in the valley bottoms are gleyi-umbric fluvisols of alluvial and colluvial origin. The soils on the interfluves tend to be slightly acidic, reaching a maximum of pH 6.7, whereas those in the valley bottoms are more acidic, with pH levels as low as 4.3. The fertility of the nitosols tend to be high owing to their depth and they are able to sustain continuous cultivation (Kefeni Kejela, 1991). In contrast, the lixosols are more fragile, with most of the nutrients and organic matter being held in the topsoil.

Ethiopia's climatic conditions are commonly classified in terms of the range of agro-climatic zones which exist throughout the country. These agro-climatic zones take into consideration the diversity of the country's altitude and rainfall, and their combined effects on the characteristics of agricultural production. Most of Illubabor falls between the agro-climatic zones of moist *Kolla* and wet *Weyna Dega* while the mountainous area to the south also represents the wet *Dega* zone (Table 4.1). This classification is largely the result of Illubabor receiving an average annual rainfall usually in excess of 2000 mm within the zone, although this varies spatially (Tafesse Asres, 1996). Lower areas such as Bure *wereda* (1550 asl) to the west and the Didessa (1600 asl) area to the east may have less than 1200 mm whereas the town of Gore (2100 asl) located in the centre of the zone has experienced annual rainfall of up to 3000 mm. The distribution of rainfall throughout the year is seasonal, with a dry season during the months of December, January and February which receive less than five per cent of the annual total. The rainfall then rises steadily between March and May (receiving 20 per cent), with nearly half of the annual rainfall being concentrated in June, July and August. The remaining 25 per cent falls between September and November (Figure 4.5).

The temperature in the zone reflects that of a warm temperate climate with a mean annual temperature of 20.7°C (Figure 4.5) (Solomon Abate, 1994), although temperatures are spatially variable throughout the zone primarily as a result of altitude. Didessa to the east of the zone experiences annual temperature variations in the range of 14.6°C to 30.4°C, whereas at Gore temperatures range between 13.3°C and 23.2°C. Temperatures also vary at different times of the year, depending on the extent of cloud cover.

Table 4.1 Agroclimatic zones of Ethiopia

Altitude	Precipitation less than 900mm	900 - 1400mm	more than 1400mm
Below 500m	**BEREHA** A: Crops only with irrigation C: No conservation S: Yellow, sandy soils T: *Acacia bussei, Commiphora erythraea*		
500 - 1500m	**Dry KOLLA** A: Sorghum, tef. C: Water retention terraces S: Yellow, sandy soils T: *Acacia*, bushes and trees	**Moist KOLLA** A: Sorghum, *nug, dagussa*, groundnut C: Widespread terracing S: Yellow, silty soils T: *Acacia, Erythrina, Cordia, Ficus*	**Wet KOLLA** A: Mango, taro, sugarcane, maize, coffee, oranges C: Frequent ditches S: Red clay soils, oxidized T: *Milicia, Cyathea, Albizea grandibracteata*
1500 - 2300m	**Dry WEYNA DEGA** A: Wheat, tef C: Widespread terracing S: Light brown to yellow soils T: *Acacia* savannah	**Moist WEYNA DEGA** A: Maize, Sorghum, tef, wheat, *nug, dagussa*, barley C: Traditional terracing S: Red-brown soils T: *Acacia, Cordia, Ficus*	**Wet WEYNA DEGA** A: Tef, maize, enset, *nug*, barley C: Widespread drainage S: Red clay soils, deeply weathered, gullies T: *Acacia, Cordia, Ficus*, bamboo
2300 - 3200m		**Moist DEGA** A: Barley, wheat and pulses C: Some traditional terracing S: Brown clay soils T: *Juniperus, Hagenia, Podocarpus*	**Wet DEGA** A: Barley, wheat, nug, pulses C: Widespread drainage ditches S: Dark brown clay soils T: *Juniperus, Hagenia, Podocarpus*, bamboo
3200 - 3700m		**Moist WURCH** A: Barley (1 crop per year) C: Drainage rare S: Black soils, degraded T: *Erica, Hypericum*	**Wet WURCH** A: Barley (2 crops per year) C: Widespread drainage ditches S: Black soils, highly degraded T: *Erica, Hypericum*
Above 3700m	Key: A: Main Crops C: Traditional Conservation S: Soils on Slopes T: Natural Trees		**High WURCH** A: None C: None S: Black soils, little disturbed T: Mountain grassland: *Artemisia, Helichrysum, Lobelia*

Source: Azene Bekele-Tesemma (1993)

58 Indigenous Management of Wetlands

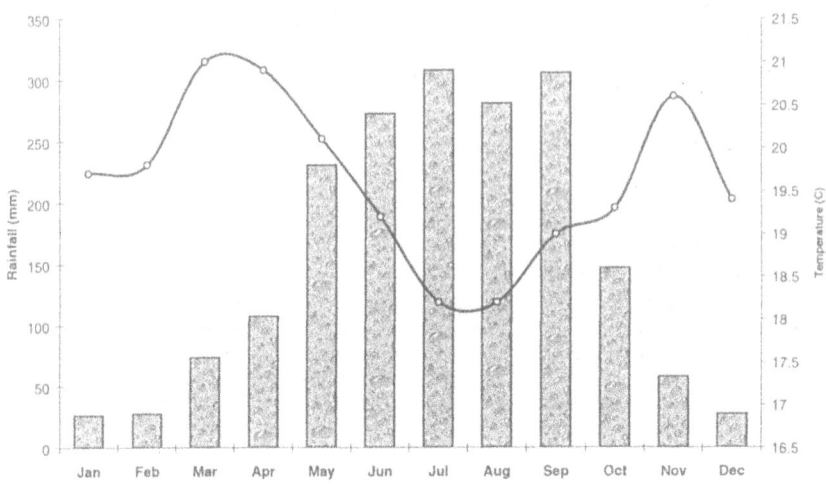

Figure 4.5 The seasonal pattern of rainfall and temperature recorded at Metu, central Illubabor (1967 – 1997)

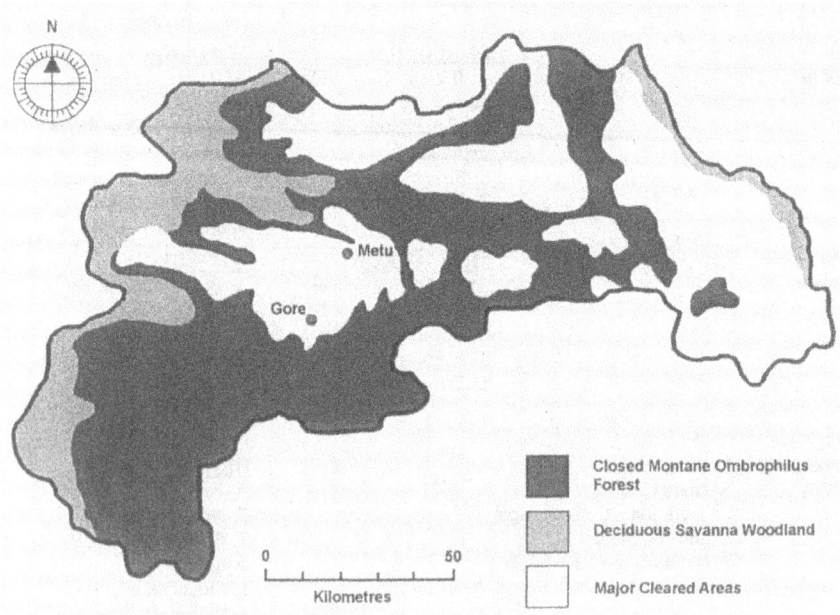

Figure 4.6 The dominant vegetation in Illubabor zone

The natural climax vegetation of most of highland Illubabor is tropical montane, evergreen rainforest where the dominant species are *Aningeria adolfi-friederici, Croton macrostachyus* and *Sapium ellipticum* (Figure 4.6). In the river valleys, which are subject to a hotter climate, *Acacia sp.* and *Olea welwitschii* are dominant. In contrast, the higher, wetter areas, located to the south of the zone, sustain dense forests of *Arundinaria alpina* (alpine bamboo) (Kumelachew Yeshitela, 1997). Vegetation in the inundated wetland areas found at the bottom of most valleys is dominated by the sedge *Cyperus latifolius*.

Demographic characteristics

The most recent government census, carried out in 1994, estimated the population of Illubabor to be 847,048 persons (CSA, 1997). Estimates for 1998 put this figure at 960,431 (of which 91 per cent live in rural areas) making Illubabor one of the most sparsely populated areas in highland Ethiopia, with an average of 58 pers/km^2 (MoA, 1998) although the population is spatially variable throughout the Zone. Estimates of population density range from 107 pers/km^2 in Metu *wereda* to 11.3 pers/km^2 in Nono *wereda*.

As a result of a history of invasion, migration and government resettlement programmes the ethnic composition of Illubabor is diverse. The dominant ethnic group, who account for 90 per cent of the population within the zone, are the 'indigenous' Oromo, who settled in the area after their invasion and expansion during the 18th century. The second largest ethnic group are the Amharas (eight per cent) who, during the late 19th century and under the leadership of Menelik II, expanded their empire to the south and the west of what was then the 'Abyssinian kingdom'. Through this conquest and a continuous process of inward migration and resettlement, the lands of present day Illubabor were subsequently annexed into what became the modern Ethiopian state by the end of the 1890s (Wood, 1977; Taye Mengistae, 1990).

Since then the most significant impact on the demographic characteristics of Illubabor has been the various resettlement schemes which were initiated during the later years of the Haile Selassie era (1968 – 1973) but to a much greater extent during the Derg government, following the famines of 1973 /1974 and 1984 (Alemneh Dejene, 1990). Resettlement, which for Illubabor meant inward migration, was seen as a means of reducing the population pressure in the northern highlands and those exposed to famine, while increasing productivity and making better use of under-utilized land (Pankhurst, 1990). According to some estimates 80,000 people from the northern areas of Wello and Tigray were resettled in Illubabor between 1985 and 1986 (Pankhurst, 1990). For a variety of reasons, many settlers did not remain in the area and returned to their own lands following the change in government in 1991.

The remaining two per cent of the population of Illubabor is, as a consequence of government resettlement, migration and immigration, composed of Tigrayan, Gurage, Mocha and Keffa peoples, with small numbers of other ethnic groups from around the country. Each of these ethnic groups have their own language, although Amharic, Ethiopia's official language, is widely spoken.

Agriculture and natural resource utilization

Much of the population of Illubabor rely on agriculture as a means of subsistence. The organization of agriculture in the area, in particular the land tenure system, has undergone several transformations since the beginning of the 20th century, all of which have had dramatic consequences for the socio-economic stability and prosperity of the area.

As a result of the colonial expansion of the Amharas, Illubabor at the turn of the 20th century was under the *gult* system of land ownership whereby the majority of land was owned by minor Amhara nobles, military or political leaders, in a system similar to the medieval European feudal system (Taye Mengistae, 1990; Bahru Zewde, 1991). Peasant farmers were tenants and landlords had rights to collect a share of the produce of all the agricultural land farmed by their tenants in exchange for the administrative and political services they provided. In turn, the landlords contributed through taxation to the government. Pausewang (1990) suggests, however, that land ownership rights up until the land reform of 1975, remained ambiguous to both the government and farmers. When Haile Selassie introduced taxation after the Italian occupation, tenant farmers were recognized as taxpayers which inferred legal ownership of their land, contrary to the opinion of the landlords.

Following the revolution and the overthrow of Haile Selassie in 1974, the land reform proclamation of 1975 attempted to nationalize and redistribute land ownership throughout Ethiopia. The private ownership of land was outlawed, tenancy was abolished and farmers were freed of the obligation to make payments to landowners. This period also saw the creation of influential peasant associations (PA) whose role included the re-allocation of farmland and the mobilization of farming activities (Brune, 1990). One of the immediate effects of the land reform was that farmers were initially able to consume more of their own produce. Faced with growing food shortages in towns, the Derg government introduced new taxes so that the newly 'affluent' farmers were forced to sell their stored excess yields and had to contribute towards the costs of administration (Pausewang, 1990). In addition, farmers and their families were also required to work for the state in a variety of communal institutions, thus diverting resources away from their own farmland.

Since the fall of the Derg regime in 1991, there have been no major changes in land access rights for farming communities. Despite the government's drive towards a free market and private enterprise, land remains 'exclusively vested in the state and in the peoples of Ethiopia' (FDRE, 1995). Although farmers do not have the right to buy or sell land, they have been assured of their inheritance rights and, in addition, crop prices are now set by the market and not through any form of state intervention.

In terms of land use, forest dominated the landscape of Illubabor during 1998, accounting for approximately 40 per cent of Illubabor's land cover. Meanwhile, cultivated and potentially cultivable land combined accounts for approximately 38 per cent of the total land area. Out of the crops grown within the zone, maize (*Zea mays*) constitutes the major cereal crop and staple food (38 per cent of crop cultivation), although at higher altitudes its growth is limited by lower temperatures. In the intermediate altitudes, maize is cultivated alongside sorghum (*Sorghum vulgare*). Another important food crop, tef (*Eragrostis tef*) is more commonly cultivated at higher altitudes, as are peas, beans and barley (MoA, 1998).

In recent years, a variety of new crops including chickpeas and sweet peppers, have been introduced by the various settlers to the area. In addition, vegetables such as potatoes, tomatoes, carrots and onions are increasingly being cultivated by farmers, with seeds and resources initially provided by the MoA. This crop diversification, characterized in some cases by intercropping, has occurred in response to the development of a more market orientated economy in the area. The most significant cash crop is coffee (*Coffea arabica*), which is endemic to the forests of south-west Ethiopia and is Ethiopia's biggest export earner. According to Solomon Abate (1994), within the Metu area of Illubabor alone, coffee occupies less than 25 per cent of the total area under crop but it generates more than 75 per cent of the annual cash income from crops. Other perennial crops include bananas, ensete, mangos and lemons but these are cultivated on a relatively small-scale. The cultivation of fast growing eucalyptus on farmland is also becoming an important source of fuelwood and a marketable commodity throughout the area.

Livestock are an important part of the farming system in Illubabor. Oxen are the main source of draught power and an important factor in crop cultivation and soil management. Grazing tends not to be restricted to certain areas of farmland except during the growing season. In many areas the practice of pen rotation is used to ensure an even application of animal manure into the soil, enhancing its productivity. Domestic animals such as sheep and goats are also reared primarily for their meat which fetches a high market price during religious festivals such as *fasika* (Easter).

Apart from its agricultural resource base, Illubabor, in contrast to the northern and central highlands, is still endowed with many of its natural resources. In particular, Illubabor zone contains approximately seven per cent of Ethiopia's natural forest resources which support a high biodiversity, and in addition, a range of timber-related economic activities throughout the zone. The climatic and topographical variation found within the zone has also resulted in an abundance of water resources which offer the potential for irrigated agriculture and the development of hydro-power through small scale barrages (TAMS-ULG, 1996, Wood, 1995; pers. comm. Hussein Jamal, 1996 (Zonal Head of MoWME, Metu)).

Water resources in Illubabor

As a result of the high rainfall, Illubabor has a rich diversity of water resources. Two major river systems in the zone, the Baro and Didessa are major tributaries of the White and Blue Nile respectively. Much of highland Illubabor is drained by upland rivers which lie at the bottom of steep, highly incised valleys. During the 1980s and 1990s, most of the zone's rivers were identified in the Baro-Akobo Master Plan Project as having the potential for a series of small hydro-electric barrages. To date, however, this potential has been exploited at only one location on the Sor River, 15 km north-west of Metu. The rivers throughout the zone are also an important source of drinking water for the urban and rural populations. During 1997, work commenced on providing a water supply for the old regional capital of Gore, which stands on an isolated volcanic plug, 2043 m asl. Until then, water was collected from numerous springs located on the sides of the mountain. Now, water is being pumped from the Keber River (1600m asl) some 12 km away.

Throughout the zone there are a variety of different types of wetlands. Floodplains are abundant and some of the larger rivers, notably that of the Geba River in the east of the zone, flood vast areas of land during the rainy season. Smaller floodplains which are usually less than 10 ha and located within the middle reaches of smaller rivers, are also abundant and widely used as grazing grounds for cattle during the dry season as well as supporting a variety of wild animal species.

In addition, seasonal and permanent swamps are ubiquitous throughout the zone in areas where runoff, streamflow or groundwater discharge is impeded and drainage conditions are poor. These conditions are predominantly found at the bottom of most valleys and these wetland areas have come to represent an important resource for the surrounding communities. These swampy areas are typical of those found throughout the high rainfall highland areas of East and Central Africa, in particular Uganda and Rwanda, where upland valley swamps are commonly used as a source of raw materials, water supply, grazing land and agriculture. Government surveys carried out in Illubabor have suggested that these wetlands, which exhibit a wide variation in their distribution and hydrological characteristics, account for approximately 1.4 per cent of the total land cover in Illubabor (MoA, 1998).

Valley bottom wetlands

The common occurrence of valley bottom wetlands in this area is a result of the high rainfall and the highly incised terrain which is host to poorly-drained depressions, facilitating the build up of sediment (Butcher and Wood, 1995). The depressions range in size from less than 10 ha to more than 300 ha, although the smaller wetlands at the heads of valleys appear to be much more abundant. Although the total percentage of land area occupied by wetlands has been estimated at 1.4 per cent (MoA, 1998), this varies in different *weredas* and ranges between 0.8 per cent in Gechi to 2.66 per cent in Yayu-Hurumu *wereda* (Table 4.2). It is important to note, however, that floodplains are not classified as wetlands but registered as grazing land in the government statistics. In addition, inaccessible wetlands located within areas of forest are also excluded from the MoA data, hence the total area of all wetlands is under-represented.

The traditional use of these valley bottom wetlands has been the collection of sedge vegetation known locally as *cheffe* (*Cyperus latifolius*). This is used as a thatching for the roofs of local houses and also as a floor covering on religious festival days. McCann (1995) presents evidence to suggest that wetlands have been

Table 4.2 The percentage of each *wereda* area occupied by wetlands

	area of wetlands (ha)	% area of wetlands	% area of cultivated wetland (1999)
Metu	2500	2.06	0.58
Ale-Didu	2448	1.47	0.33
Halu-Bure	1513	0.96	0.33
Yayu	3600	2.56	0.19
Alge-Sachi	1280	1.45	1.37
Darimu	2102	1.47	0.19
Chora	2056	2.34	0.46
Dega	1080	0.99	0.29
Bedele	1721	0.80	0.25
Gechi	1160	0.97	0.54
Dedessa	887	1.07	0.20
Sale-Nono	2330	1.01	unknown
Illubabor Total	22677	1.36	

Source: MoA (1999).

part of the agricultural system in the highland forests of the south-west since at least the 1880s, when they were used for maize cultivation. The Italian geographer Antonio Gecchi, who recorded this information during a visit to the south-west, also provides the first description of these areas:

> The end of this valley presents an extended plain, little cultivated although the soil was very fertile and rich in water, which for lack of slope here and there, was waterlogged. The dry part forms a luxurious meadowland where beautiful cattle and sheep grazed (McCann, 1995, p.163).

Recorded information on the nature, abundance and utility of these valley bottom wetlands in Illubabor is scarce although they have recently been the subject of development initiatives by the MoA and the NGO Menschen für Menschen (MFM). A report investigating the potential of valley bottom sites for agriculture by MFM (1995) provides some basic information on their hydrology. In this report, the source of water for valley bottom wetlands is identified as 'overflow from adjacent stream channels', 'runoff from the catchment' and an 'increase in the level of the water table'. Valley bottoms are also classified into two morphological types, namely *open* and *closed* formations, where *open* refers to those wetlands distinguished by a visible outflow and those without are *closed*. Tafesse Asres (1996) suggests that the soils in these valley bottoms are mainly gleyisols consisting of recent alluvial or fine colluvial deposits and showing hydromorphic properties. The texture of the soils varies from silt loam to silty clay loam with an increase in clay from a depth of 20 cm. Valley bottom soils also tend to be weakly acidic.

It has been suggested that these valley bottom wetlands perform important hydrological functions which include the storage and regulation of water after rainfall events, and the trapping of sediment transported from the surrounding catchment (pers. comm. Solomon Abate, 1996). In addition, the role of *Cyperus latifolius* in water purification has been reported by Jahn (1981), who gives an account of its association with uncontaminated drinking water by local communities in Sudan. Throughout Illubabor, farming communities also rely on these wetlands for the provision of water for their own consumption (from the peripheral springs) and for their farm animals, particularly during the dry season.

As well as *Cyperus latifolius* (which is harvested as a roofing material and for floor covering), other wetland plants such as palms (*Phonex reclinata*) are reportedly used as handicraft materials and for medicinal purposes (Afework Hailu, 1998). For the provision of all these natural functions, valley bottom wetlands are clearly an important resource to local communities, yet throughout Illubabor their agricultural potential has also been regarded as important, owing to their rich soils and abundant water supply.

Valley bottom development

As a result of their association with liverfluke and mosquitoes, Illubabor's valley bottom wetlands have generally been avoided by people and their livestock (Solomon Abate, 1994). With the expansion of settlement down the slopes of the valleys, there has been some limited use of wetland margins for the cultivation of green maize, which helps meet the food needs in the 'hungry season' (May – July), before the main harvest (Wood, 1996). In some areas of Illubabor farmers have extended this activity further into the wetland (Tafesse Asres, 1996), with the construction of simple drainage channels over an area which is usually less than 0.5 ha at the edge of the wetland.

The maize crop is grown during the last part of the dry season and the first part of the rains (February – June) using a combination of groundwater and rainfall (Butcher and Wood, 1995). After the green maize is harvested, the sites are abandoned for eight months and the wetland vegetation re-establishes itself as the drainage channels become silted and blocked. Used this way, the cultivation of maize in the valley bottom appears to be sustainable, presenting no long-term problems in terms of crop yield or hydrological and ecological changes (Wood, 1996).

Recently, however, there has been concern over the increasingly intensive use of whole valley bottom wetlands for multi-cropping. Rather than limiting drainage and cultivation to small plots within wetlands, many wetlands have typically undergone complete drainage which involves the excavation of drainage channels up to one metre in depth, in a 'herringbone' layout throughout the whole wetland. This facilitates the cultivation of maize (which is often intercropped with cabbage) throughout the wetland.

Some farmers have, in addition, begun to practice crop cultivation in the wetland immediately following the rainy season (September) and prior to the main maize crop. This typically involves cultivating tef on the residual moisture, which matures within three months and is harvested in late November or early December (McCann, 1995; Tafesse Asres, 1996). Following this harvest, the wetland is prepared for re-cultivation with maize, starting in late December or early January. In some areas of Illubabor where this intensive form of wetland drainage and cultivation is taking place, farmers have identified a range of problems which include a fall in wetland water table levels and the failure of valley bottoms to support dry season agriculture after a number of years of drainage (Wood, 1995; Afework Hailu, 1998). Many wetlands throughout Illubabor (and in neighbouring zones) which are now only suitable for rough grazing are reported to have undergone a process of agricultural intensification to the point where production ultimately became unsustainable. It has been suggested that a general trend towards the intensification of valley bottom agriculture, with unintentional impacts on wetlands, may have been influenced by five inter-related factors:

The promotion and expansion of coffee production Illubabor has become a dominant area in the production of coffee since the 1960s and as a result, has been the subject of investment in its road network (for marketing reasons) and in terms of agricultural extension projects. As successive governments recognized the area's potential for coffee production, more areas of forest land surrounding settlements have been converted for coffee growing. These areas would have otherwise been used for new arable land, hence the need to develop new areas for crop cultivation. The result has been a movement into valley bottom agriculture (Tegegne Sishaw, 1998).

Regional food self sufficiency Following the 1984 drought, the Derg government issued a directive requiring all regions to become self sufficient in terms of staple food production. The effect of this was the formulation of wetland development policies by the MoA and the Ministry of Tea and Coffee Development (MoTCD). In *weredas* classed as major coffee growing areas, the forest was retained for increased coffee production while wetlands were regarded as principal areas for expanding crop production. In other *weredas* wetland cultivation was also encouraged in order to achieve a state of food security, particularly through the early harvesting of green maize (Afework Hailu, 1998).

Resettlement The arrival of approximately 80,000 settlers in Illubabor from the north of Ethiopia occurred between 1984 and 1989 primarily as result of the government resettlement programme which was set up in response to the 1984 – 1985 famine (Pankhurst, 1990). As a result, Illubabor experienced population pressure and a shortage of arable land for the settlers to cultivate. Although forest clearance provided a means of addressing the land shortage problem, this was not always possible and it conflicted with the government's strategy of increasing coffee production (see above). Instead, settlers were allocated marginal areas such as the forest fringes (undesirable owing to the abundance of wild pests) and valley bottom wetlands, which were brought into production and their use encouraged by the MoTCD.

The integration of settlers into local communities and the redistribution of existing farmland on the uplands between settlers and the 'indigenous' population, also stimulated competition and rivalry between ethnic groups for the better farming resources (Alemneh Dejene, 1990). Faced with shortages of productive farming land, many settlers initiated the cultivation of valley bottom wetlands which offered significant advantages in terms of available moisture and soil fertility.

Villagization During 1986 and 1987 the government adopted a policy of villagization in an attempt to relocate farmers who lived in scattered farmsteads to a few centralized villages in each *kebele*. This policy was, however, largely unsuccessful and unpopular with farmers who, after cultivating land around the new villages by day, walked back to their own farms at night. Whilst this policy had the effect of reducing wetland cultivation around old farmsteads, the use of those wetlands located around new villages was intensified (Afework Hailu, 1998).

Market economy Finally, with the development of a market economy in the area, many farmers have regarded valley bottom wetlands as a valuable resource for increasing their crop yield beyond the subsistence level. Crops such as Irish potatoes, carrots and cabbages are now being farmed specifically for the marketplace, while some 20 per cent of the maize grown in wetlands is also being sold (Solomon Mulugeta, 1999).

Government policy directly affecting wetlands

In Illubabor both the MoA and the Ministry of Water, Minerals and Energy (MoWME) have a vested interest in valley bottom wetlands. The MoWME was set up in 1988 although a zonal office in Metu has only existed since 1995. In Illubabor zone, the brief given to this Ministry by central government was to develop irrigation schemes and river diversion, with the aim of increasing the amount of agriculturally productive land. In reality, the topography and physical characteristics of Illubabor have prevented the widespread development of irrigation schemes and the MoWME has turned its attention instead, towards the drainage of wetland areas. This policy is also motivated by the need for increased food production, both for the market place and to alleviate food shortages in an increasing population. The MoWME does recognize that problems may exist following wetland development, but these are generally limited to concerns over reduced river discharges with knock-on effects for hydro-electric schemes.

The role of the MoA has been to encourage the drainage and cultivation of wetlands as part of its overall policy of increasing agricultural production in the zone. There are no technical staff who deal solely with wetland utilization and the only guidelines for wetland cultivation in existence are those developed by the Natural Disaster Prevention Committee in 1988 (Table 4.3). It is the role of other technical staff to provide advice on wetlands where their own specific discipline, e.g. horticulture, soils or irrigation, relates to wetlands (pers. comm. Afework Hailu, 1999). Nonetheless, as a result of the various extension activities disseminated by development agents, wetland development tends to be encouraged. The degree to which different areas receive assistance appears to differ throughout the zone and at the *kebele* level. In some areas farmers have been targeted for assistance and they report wetland development initiatives occurring over the last 15 years. In other areas farmers state that they have had no input from the MoA with regard to managing their wetland plots (this situation is examined further in Chapter Eight).

Table 4.3 Guidelines for wetland development as established by the Natural Disaster Prevention Committee (1988)

1. Main drainage channels that are to be constructed to drain wetlands should be 50 metres away from the head of wetland and water springs.

2. After the construction of drainage channels, reeds should be cut and used for another purpose if the wetland to be developed is pristine.

3. As soon as the drainage channel is dug and the reeds are cut, cultivation should start immediately and the field should be sown as soon as possible. This is to conserve the moisture in the soil for germination and crop growth.

4. The crop field should be kept weed free in order to reduce moisture competition.

5. If the wetland is a major water source of a river it should not be cultivated since it will have an impact on downstream users of the river.

6. If the wetland to be cultivated was allocated for grazing before, then evaluation of the benefits from grazing and cultivation should be assessed and priority should be given to the one with the highest benefit for the community.

7. Crops to be grown should be those that mature before the main rainfall season starts.

Source: Afework Hailu (1998)

One further influence on wetland development activity has been the zonal and *wereda* administration. During the Derg regime, although having no clearly defined policy on wetland utilization, the administration had the authority to enforce government policy upon the various government departments. Hence, where the *wereda* agricultural office was seen as not significantly reducing food shortages, the political administration pressured the MoA to bring the 'marginal' unused wetland areas into agricultural production. At present, administrators at the *kebele* level actively encourage farmers to drain wetlands, while also arbitrating in any disputes which arise through wetland utilization (Afework Hailu, 1998).

The role of NGOs in valley bottom development

The most influential NGO in Illubabor has been Menschen für Menschen (MFM) which was initially involved in the provision of food and farming equipment for the resettled farmers. Since the start of their involvement in 1985, these activities have been expanded to include the provision of schools, health clinics and water supplies and, more recently, addressing the environmental impacts of the settlers. In 1990 this concern led to the creation of an 'Eco-development Project' which aimed to develop sustainable agricultural production and reduce the level of forest clearance which had increased dramatically since the settlers arrived. One of the specific aims of the project was the development of valley bottom wetlands, with irrigation and drainage to improve crop yields and food security (Butcher and Wood, 1995). MFM's role was the provision of technical assistance in surveying and ditch excavation and the provision of hand tools and vegetable seeds. In 1995 this was targeted towards 81 ha of valley bottom wetlands spread throughout Illubabor zone (MFM, 1995).

According to MFM the development of these sites would:

1. provide 2 – 3 crops a year;
2. produce an early maize harvest to address the hungry season before the main harvest;
3. ensure more reliable harvests because of the use of groundwater and irrigation;
4. reduce the rate of deforestation by providing an alternative source of land for cultivation (MFM, 1995).

Drainage systems introduced under the valley bottom development programme have varied in their construction and have generally been improved upon over time as farmers' sensitivity to local conditions has improved. Early attempts involved creating channels one metre deep which led to over-drainage of the wetlands. More recently, channels of 40 cm and 60 cm in depth have been used. It has been estimated that during the first five years of the project over 300 ha of valley bottom wetlands have been developed with assistance from MFM (Wood, 1996), although this accounts for a fraction of the total area of wetland under development in Illubabor (Figure 4.7).

The recent concern among local communities over the consequences of valley bottom drainage coincides with some problems experienced by MFM's valley bottom development programme, notably in their demonstration site near the village of Hurumu (Figures 4.8 and 4.9). This valley bottom wetland was first drained and cultivated by MFM in 1993 (although having undergone indigenous development in the past), providing a showcase for new agricultural initiatives which included the multi-cropping of vegetables and an attempt at fish farming. Figure 4.8 shows this wetland under cultivation in 1994 and Figure 4.9 (1997) after the site had been abandoned for reasons which remain ambiguous. What is clear, is that the wetland has been altered significantly by the drainage and cultivation process, which in this case is characterized by wide and deep drainage channels. Since abandonment, these drainage channels have been replaced by large gullies and eroding banks, with a water table well below the surface of the wetland. Furthermore, the surface of the wetland has developed into a dry grassland ecosystem with an obvious loss of natural wetland vegetation. In this state the wetland's utility is confined to supporting cattle as a rough grazing area, as would appear to be the common trend in several parts of Illubabor, especially Bedele *wereda* where there are reports of a serious shortage of *cheffe* following widespread wetland cultivation (pers. comm. Afework Hailu, 1998).

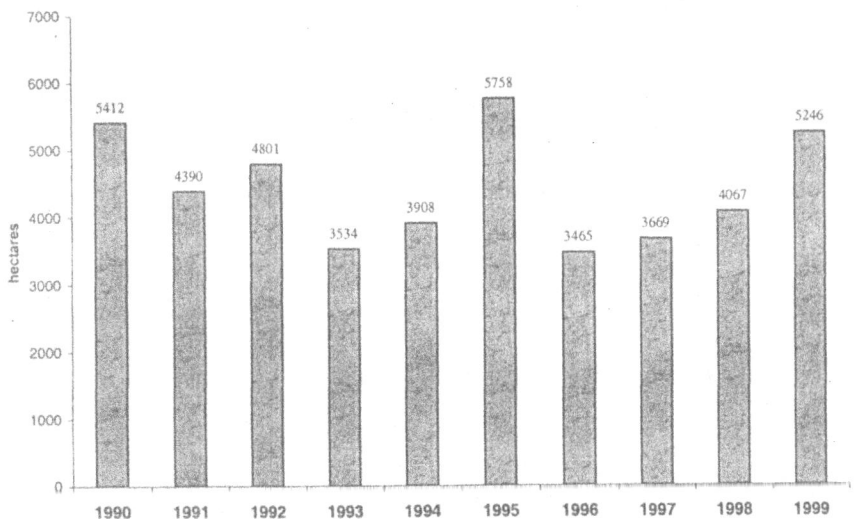

Figure 4.7 **Estimates of the total area of wetland under cultivation (ha) in Illubabor zone since 1990**

Source: MoA (1998, 1999)

MFM have become aware of these problems and have established several guidelines to be followed in their future wetland development and activities (pers. comm. Habtamu Wubshet, 1996 (Eco-development head)). These have included:

1. retaining wetland areas at the head of valley bottoms so that the ratio of swamp to cultivated land is not less than 1:10;
2. the maintenance of forested headwater catchments;
3. the adjustment of drainage channel depth to prevent over-drainage.

Figure 4.8　Hurumu wetland under cultivation in 1994

Figure 4.9　Hurumu wetland in a degraded state during 1997

Butcher and Wood (1995) raised further questions about the use of valley bottoms which relate to reductions in soil fertility, water table levels and the overall impact on catchment hydrology. The problems associated with wetland development will also have an impact on the socio-economic situation for both the individuals and communities, whose extension into valley bottom cultivation has led to the commodification of farming in the area. The effects of hydrological disturbance on the biodiversity of the wetland ecosystem could also be significant, leading to a loss of an important wildlife habitat.

The problem remains that there is little information on the functioning of valley bottom wetlands in the region. In addition, one of the main criticisms of MFM's valley bottom development programme has been that no monitoring has taken place and that this has led to short-sighted development initiatives with little thought for their consequences.

The research in context

In summary, valley bottom wetlands are ubiquitous throughout most of Illubabor zone, where they are an important natural resource. Traditionally they have been used by local communities as a source of roofing material for their houses, drinking water and in some cases for medicinal plants. There is also some evidence to suggest that these wetlands have been farmed on a small-scale in the past in order to bridge the food shortage gap between the hungry season. More recently, wetland cultivation appears to have been expanding and intensifying as a result of a combination of population pressure, socio-economic changes or political interventions. At the same time, there have been reports that farmers are experiencing problems with their wetland cultivation system, problems which include a decline in fertility and a drying out of the wetland soil.

The evidence suggests that Illubabor's wetlands have undergone a typical shift from a sustainable form of utilization to an unsustainable system of intensive use characterized by natural resource degradation. As discussed in Chapter Three, this is a common occurrence in many parts of the developing world where increasing external pressure on local communities results in natural resource utilization at an unsustainable level. In particular, the case of swamp cultivation in Uganda, highlighted by Denny and Turyatunga (1992) provides a relevant example of the effects of agricultural intensification and the shift to an unsustainable mode of wetland utilization.

The specific situation of Illubabor's wetlands, however, is represented in Figure 4.10, which illustrates the hypothetical shift from sustainable to unsustainable utilization. Figure 4.10a represents the situation in a 'traditional' system of wetland farming where drainage and cultivation is taking place but based on an indigenous system of management. In contrast, Figure 4.10b shows the same wetland system under conditions similar to those described in this chapter with respect to Illubabor's wetlands. In this model, external factors play a critical role in placing pressure on the wetland system (or on the farmer to develop the wetland) resulting in agricultural intensification.

Indigenous Management of Wetlands

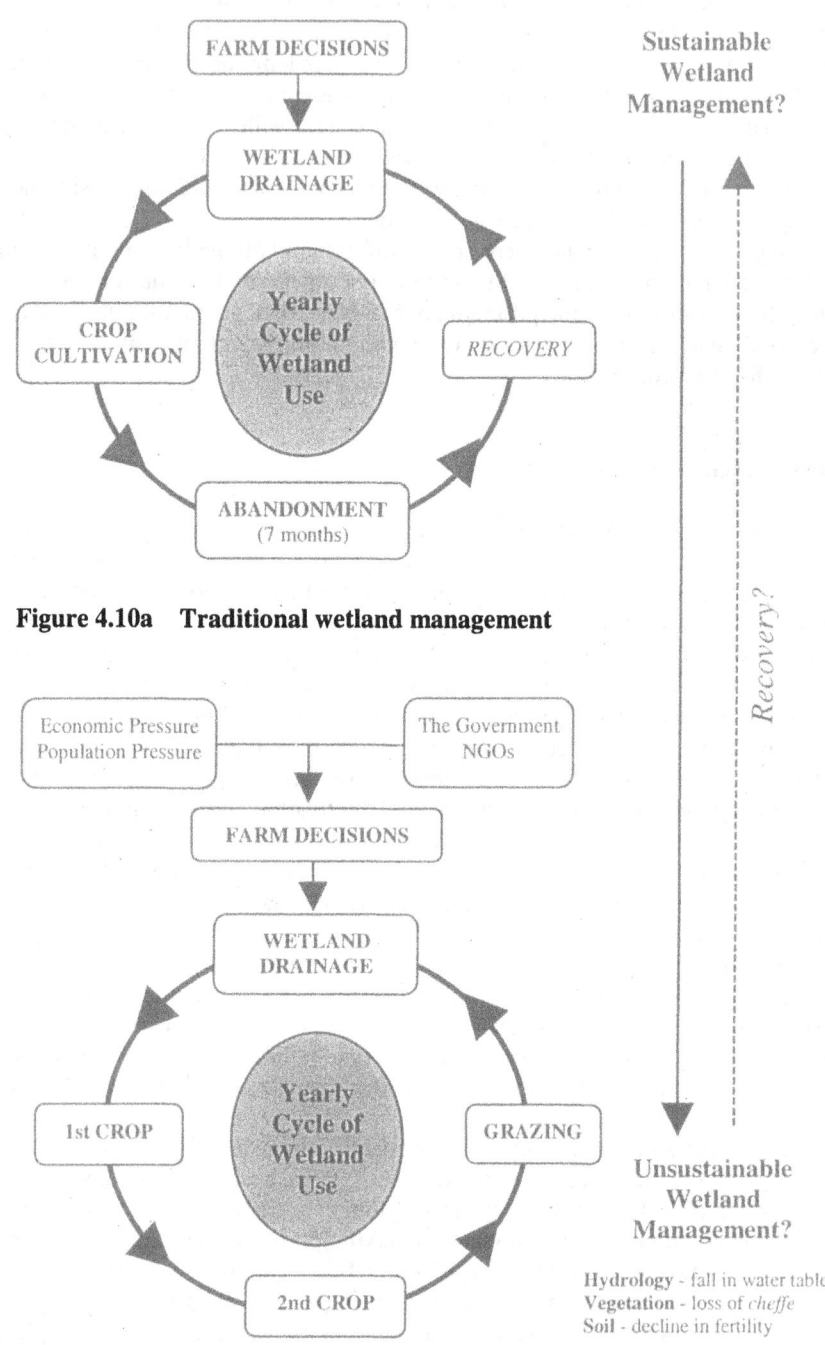

Figure 4.10a Traditional wetland management

Figure 4.10b The intensification of wetland utilization

This model also represents the situation on which the research presented in this book is based. To reiterate the aims stated in Chapter One, the research sought to determine the extent to which the hydrological management of Illubabor's wetlands for agricultural development could be sustainable. This was undertaken through pursuing four main objectives:

5. the identification of the general characteristics of valley bottom wetlands found in Illubabor Zone;
6. the identification of the impact of agricultural utilization on the hydrological regimes of these valley bottom wetlands;
7. an exploration of the role of indigenous knowledge in wetland resource management and the ways in which wetland knowledge held by local communities is operationalized;
8. the identification of principles for the sustainable hydrological management of Illubabor's valley bottom wetlands.

While Objective Three represents the outcome of the research, Objectives Two and Three address the mechanisms of this shift from sustainable to unsustainable utilization.

Objective Two specifically addresses the question of sustainability from the hydrological point of view and a major part of the research was aimed at establishing the characteristics of wetland hydrology under those conditions represented in Figure 4.10a and 4.10b.

Similarly, Objective Three represents the need to understand the precise ways in which local communities within Illubabor interact with wetlands and the hydrological management process, and the mechanics of their indigenous wetland knowledge system, particularly within the context of the two scenarios in Figure 4.10.

The first stage of the research process, however, centred on the fulfilment of Objective One, enabling a more accurate picture of Illubabor's wetland resources to be built up prior to the detailed hydrological and IK investigations. The next chapter, which focuses on the methodology employed during the research, starts by highlighting the classification process from which wetland types were identified.

Chapter 5

The Research Approach

Introduction

This chapter describes the principal methods used during the research, presenting the methodological steps taken in order to achieve a representative sample of wetlands from the range of different wetland characteristics found within highland Illubabor. It then discusses the range of field methods employed in the implementation of the hydrological monitoring programme and secondly, the investigations of indigenous wetland utilization.

The study area

Illubabor zone, comprising an area of approximately 16,555 km^2, was deemed too large and diverse a terrain in which to carry out a programme of hydrological investigation on valley bottom wetlands on a practical and efficient basis. For this reason a specific representative field study area was chosen following preliminary visits throughout the zone. The study area chosen comprised an area of approximately 3025 km^2 and is represented by four 1:50 000 scale topographical sheets from the Ethiopian Mapping Authority, dated 1986 (Figure 5.1).

This area was considered suitable for several reasons:

1. over 100 valley bottom wetlands of different characteristics are located within the area. This may be linked to the topography of this area which is predominantly composed of steep sided, sharply dissected valleys with dendritic drainage patterns;
2. it represents an area in which valley bottom development has taken place for a number of years. Many of the cases of wetland development appear to have been spontaneous or through the influence of MFM and the MoA and some have documented case histories;
3. the town of Metu is the zonal administrative capital and houses the MoA, MoWME and MFM, all of which are sources of information on the current and historical utilization of valley bottom wetlands;
4. most of the valley bottom wetlands within the area are relatively accessible by road and well within a day's journey of Metu.

76   *Indigenous Management of Wetlands*

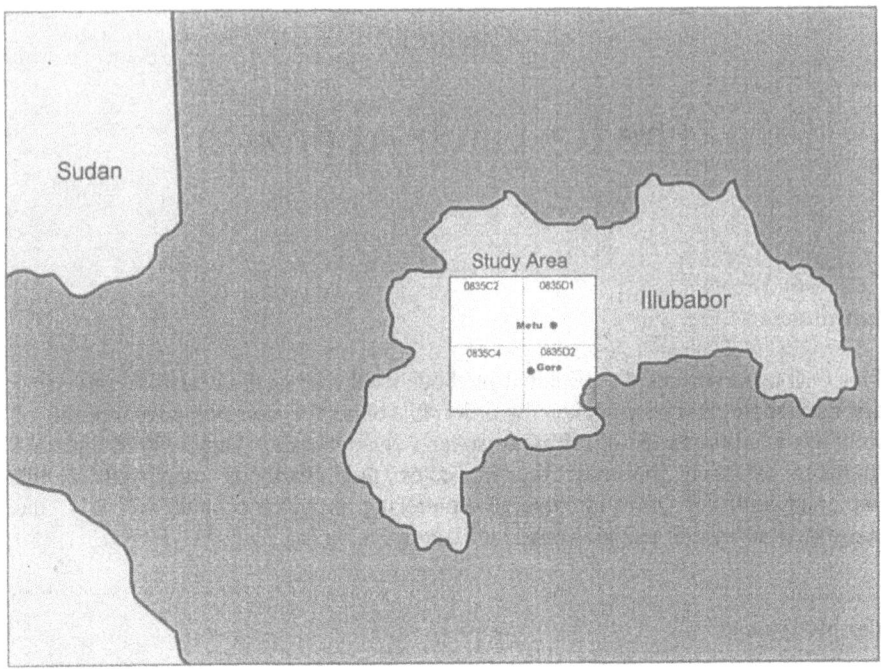

Figure 5.1   The location of the study area within Illubabor

The first field visit to the study area took place between February and May 1996, during which information on the hydrology, geomorphology and land use of a variety of different wetlands was collected. Simple measurements of water quality were taken as well as descriptions of soil characteristics, vegetation and farming practices, in order to identify the range of variation between wetlands. With the aid of an interpreter, information on the history of wetland use, population dynamics and land use changes were also elicited from local farmers. Several meetings were also held with officials from the MoA, MFM and the MoWME.

As a result a small database of wetland sites was established, with varying amounts of data on the physical, social and historical characteristics of each. In summary, these initial investigations suggested that:

1. wetlands tend to have similarities in their natural vegetation although there is considerable variation in their hydrological characteristics. Water sources, drainage channels, stream discharge and soil moisture among others, were found to be extremely variable. This could be explained by the variation in the size of wetlands throughout the study area and the history of their use;
2. variation in the use of wetlands is evident. Some were being used, others remained undisturbed but the reasons for these variations are not clear;
3. degradation, characterized by changes in wetland hydrology and soil, does

seem to occur following a period of development;
4. according to some farmers, wetlands do recover from this damaged state;
5. wetlands appear to be divided into individual plots distributed amongst the community's farmers;
6. the most common use of wetlands appears to be the cultivation of maize and cattle grazing;
7. most of the farmers claim that the fertility of the wetland declines following drainage.

The initial field visit made it clear that the subsequent methodology of study site selection had to accommodate the variability of the wetlands, whilst also addressing the aims of the research. To reduce the variability and in order to focus on any relationships between different wetlands, a broad classification of all the sites within the study area was undertaken.

Wetland classification and site selection

The classification of wetlands is a fundamental step in any research which considers their management or conservation. For the purposes of this research, the classification of wetlands into different types was undertaken primarily to identify the variety of physical characteristics, which potentially influence the hydrological regimes and subsequently the human utilization of these areas. From the resulting typology, a representative sample of valley bottom wetlands were selected for more intensive field study. Figure 5.2 summarizes this classification process, the results of which are presented in Chapter Six.

Wetland variability in the study area

The first phase of the classification process involved the identification of all the valley bottom wetlands in the study area using the 1:50 000 scale maps. The key to these maps illustrates two forms of wetland which are present in the area: '*marsh*' and '*impassable swamp*' and analysis of these maps revealed the presence of 114 of these wetlands, which were digitized and entered into a Geographical Information System (GIS). This provided a simple way of producing a database of wetlands as well as the means to calculate variables such as the area of each wetland. The information held in the GIS included data for wetland size, wetland order (the number of wetlands preceding a particular wetland in the catchment), the inflow – outflow characteristics of each wetland (in terms of visible channels), wetland shape (as a ratio of perimeter to area) and the presence of cultivation which was determined through the analysis of aerial photographs.

Having constructed a database of wetlands in the area, the next stage in the classification process involved analyzing the records and classifying wetlands into types on the basis of the measured variables. This was undertaken using descriptive statistics and manipulating data on the GIS. The wetland types identified were then

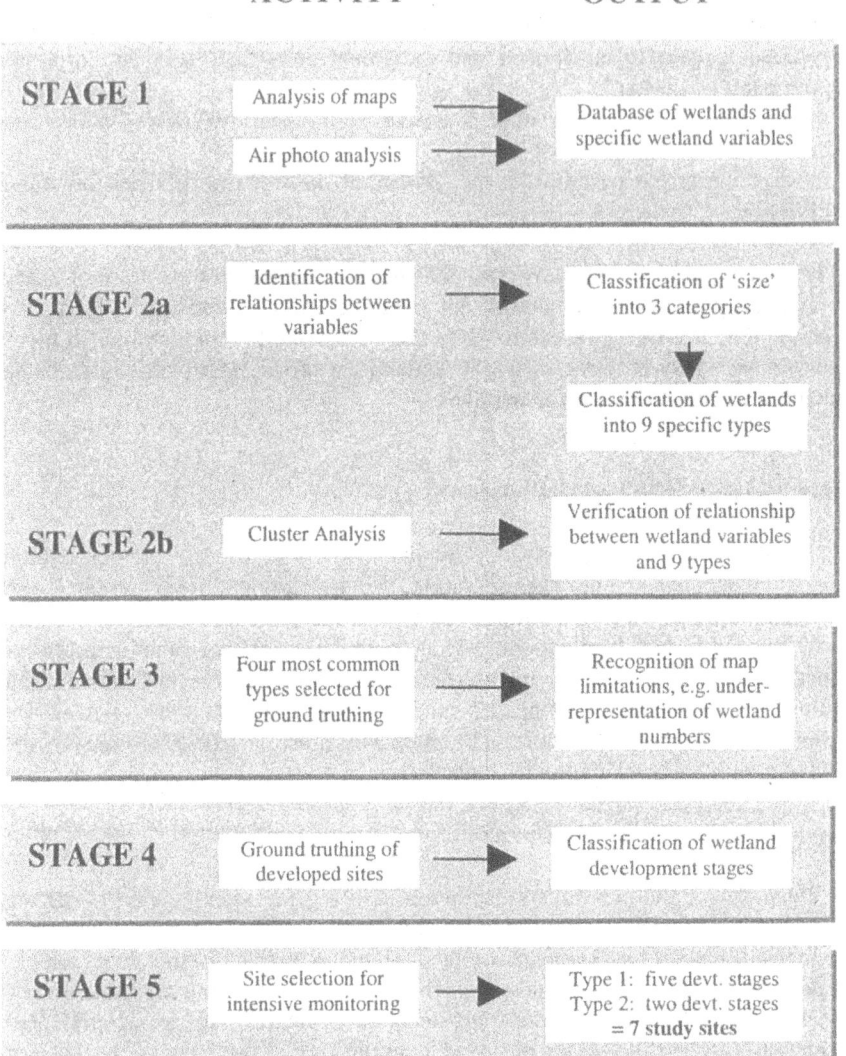

Figure 5.2   Summary of the classification and study site selection process

verified using the cluster analysis technique which identified clusters of similar wetlands on the basis of the strength of associations between the different measured variables.

Ground truthing

The typology of wetlands formed the basis of the site selection criteria, where wetlands were selected according to how representative they were of those types in the study area. Thirty sites were chosen from the map of the study area to represent the four most commonly occurring wetland types which would be visited during the early stages of fieldwork. The aim of the site visits was to supplement the existing data derived from the GIS with more information on the wetlands' physical catchment characteristics e.g. soil type, geomorphological characteristics and the duration of flooding. In addition, the site visits constituted a ground truthing exercise which would highlight any inaccuracies in the data extrapolated from the maps.

The classification of wetland development

The methodology for the selection of specific wetlands was based upon the ergodic hypothesis of space-time substitution (Summerfield, 1991), which proposes that sampling in space is the equivalent of sampling through time, providing that the distribution of objects or events over space and through time are the same. If this is the case, sampling is interchangeable through time and space. Applied to the context of wetlands, this hypothesis proposes that the temporal variation in wetlands as a result of drainage is also expressed in the spatial variation of sites throughout the study area. The study sites selected were, therefore, representative of the different stages in their hypothesized development cycle.

During the ground truthing, information on wetland use was collected through observation and short, informal discussions with local farmers at the 30 sites visited. This included information on the types of drainage taking place, the history and length of the cultivation period, and the actual area under cultivation. This led to an overall representation of the different uses and stages of valley bottom wetlands which was then used in formulating a conceptual model of wetland development (Figure 6.7). This model formed the basis of one criteria for site selection – that of the temporally variable development status of the wetland.

Site selection criteria

By combining the information from the typology with that of the land use history of each wetland, it was possible to comment on the predisposition of specific types of wetland to cultivation and development activities. This was the main consideration in choosing specific sites for intensive monitoring. Those selected would be representative of the most common types undergoing development.

The second criteria for site selection was the representativeness of the wetland landform itself, which relates back to the typology. Those wetlands selected would:

1. be representative of the most common types of wetland in the study area;
2. represent those wetland types which have undergone the most intensive development and which are consequently in need of sustainable management;
3. be representative of wetlands and the issues surrounding wetland utilization, as reported in the relevant literature;
4. be both accessible and suitable for data collection.

The selection of two wetland types was judged to be both practical and representative, providing an opportunity for analysis of the variation between and within these specific types.

With the selection of study sites established, two main approaches were used to pursue the aims of the research. First, a hydrological monitoring regime was set up to address the question of the effects of human interference on the wetlands hydrological regime. Secondly, a framework for analyzing and investigating the role of indigenous knowledge in wetland utilization was developed.

Hydrological monitoring

The monitoring of wetland hydrology is a fundamental part of the process of building up an understanding of any wetland system. The main focus of this research was the groundwater and soil moisture storage components of the wetland, and the interaction of this with surface water and rainfall. Consequently, the hydrological monitoring regime incorporated investigations into the water table elevation, water quality, saturated hydraulic conductivity of wetland soils and rainfall.

Water table measurement

In the wetland hydrological system, flows between individual storage components tend to be slow and are difficult to measure directly. They do induce, however, measurable responses in the water table elevation, which if monitored, can provide a means of assessing the hydrological behaviour of a wetland site (Gilman, 1994). Consequently, the monitoring of the water table on all the study sites formed the core of the hydrological monitoring programme. This addressed the need for base line information on the hydrological behaviour of natural, undisturbed wetlands, and also the hydrological changes which take place as a result of drainage and cultivation.

Out of a range of methods available for measuring the water table in wetland soils the dipwell method described by Bouma *et al.*, 1980 and Baird and Ross, 1992 was used throughout the research on account of it combining simplicity, robustness and accuracy. Dipwells (Figure 5.3) were constructed from PVC pipes (approximately 2 cm in diameter and 1 metre in length), with approximately 20 perforations of 0.5 cm diameter drilled along their length at 10 cm intervals. These were inserted into the floor of the wetland so that the base of the dipwell was positioned below the normal low water level and the top just above the ground

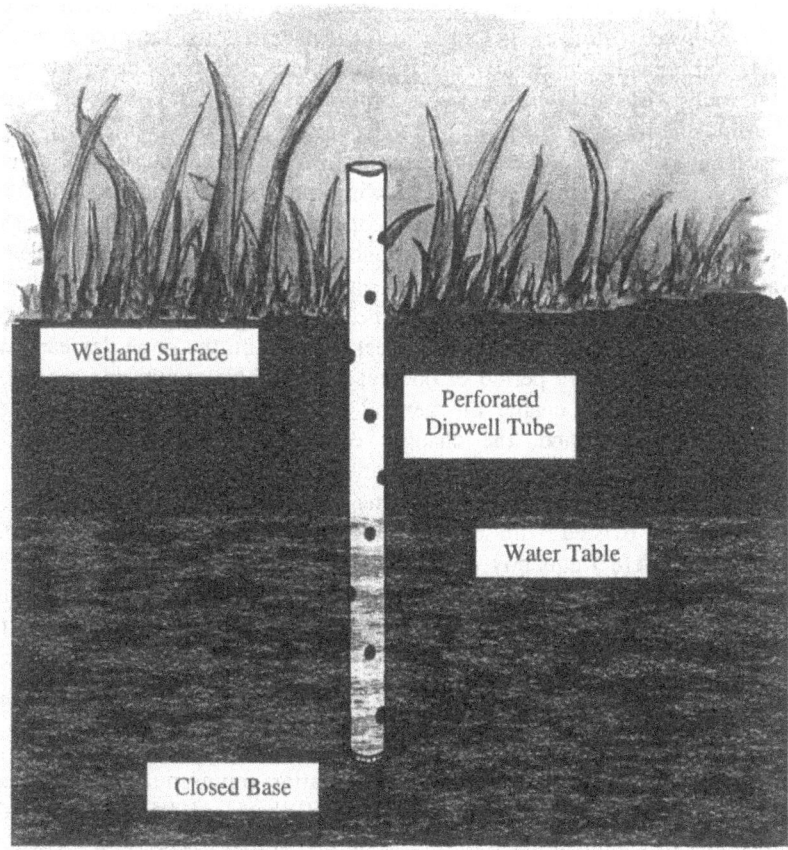

Figure 5.3 The dipwell apparatus

surface and the expected rise of the water table.

Between nine and twelve dipwells were installed in the headwater area of each study wetland depending on the size of each wetland but also upon the requests of the various wetland owners. The headwater area of each wetland was chosen for the dipwell transects on the basis that these areas receive more hydrological input (runoff) than those further down the wetland, therefore representing key farming areas in terms of their soil moisture regime. Furthermore, research on *dambos* suggest that these upper areas retain moisture throughout the year whilst those towards the outflow of the wetland may dry up completely under drought conditions (Turner 1986; Faulkner and Lambert, 1991).

The dipwells were arranged in latitudinal and longitudinal transects across the valley bottom, a technique that has been reported to provide an adequate assessment of water table fluctuations (Wheeler and Shaw, 1995). By setting dipwells in transects, the number of dipwells used to represent the spatial variation of the water table can be minimized and the data obtained can be represented easily in graphical form. Furthermore, dipwells located in transects are much easier to find than those arranged in an irregular pattern, an important practical consideration for valley bottom wetlands where cultivated maize can grow up to a height of four metres and where *Cyperus latifolius* is often in excess of 1.5 metres.

The dipwells were measured using the technique proposed by Reeve (1986) in which a length of graduated plastic tubing is lowered down the dipwell tube and contact with the water is discerned by blowing into the tube and listening for the sound of bubbles. At this point, the height of the plastic tube is measured against the top of the dipwell, giving the total distance between the top of the dipwell and the water table elevation. The water table height is then established by subtracting the height of the dipwell protruding above the surface from the first measurement.

Dipwell records were collected on a weekly basis, on the same day each week whenever possible, over a period of one year. Under this monitoring regime, the data were sufficient to enable the general water table characteristics of each wetland to be established, the changes occurring throughout the year and significantly, the variation within each wetland to be identified.

Throughout the hydrological monitoring programme it was recognized that several problems existed with respect to dipwells and their measurement. In particular, they were frequently destroyed or uprooted by humans and animals, necessitating their replacement and it was not possible on all occasions to ensure that replacement dipwells were inserted in an identical location to their predecessor.

Water quality measurement

Drainage and subsequent agricultural intensification represents one of the most extreme forms of damage to wetland ecosystems (Wheeler, 1995). This can induce a wide variety of chemical changes in the soil and water, related to a reduction in water table levels. In the nitrogen cycle, oxidation of ammonia caused by exposure to air, results in the formation of nitrates which are leached out of the soil or taken up as fertilizer by plants (Hill, 1976). In organic soils, drainage aerates the soil which initiates chemical and biological transformations resulting in the release of solutes. The release of solutes by wetland soils can be easily measured in the drainage network and provide some indication of the nutrient status of the wetland and the changes taking place (Naucke *et al.*, 1993).

The measurement of both pH and electrical conductivity can give a general indication of a wetland's chemical environment. The pH provides a measure of the acidity and alkalinity of water, which may be affected by the leaching of nutrients from the wetland soil and electrical conductivity is a measure of the concentration of dissolved salts (Rowell, 1994). Both parameters are affected by dissolved substances in the water, some of which may originate from outside the wetland. For this reason, the monitoring of pH and electrical conductivity is an effective way of identifying the balance of nutrients within a wetland, in terms of nutrient levels in both the inflow and outflow.

Nitrogen and phosphorous are principally indicators of soil fertility (Brooks and Stoneman, 1997). They are taken up by wetland vegetation and either stored or transformed into microbial or plant matter. The decay of organic material (as a result of oxidation) releases nitrogen and phosphorous in soluble organic compounds and depending upon the hydraulic residence time, these can be released from the wetland soil (Horne and Goldman, 1994). Through the monitoring of nitrate and phosphate concentrations in the hydrological network of wetlands under different development conditions, the research sought to establish the influence of drainage on wetland nutrient flows. In addition, by monitoring nitrate and phosphate concentrations throughout the year in association with other hydrological parameters, the relationship between seasonal changes in nutrient concentrations and the wetland hydrological regime could also be established.

For each of the seven study sites, the monitoring of pH and electrical conductivity was undertaken on a weekly basis and nitrate and phosphate on a monthly basis. Two sites within each wetland were monitored: the head of the wetland, which consisted of either an inflow channel or a spring, and the outflow of the transected area. The pH and electrical conductivity readings were taken using a small, temperature compensating, hand held meter.

All samples were analyzed using a Camlab DR200 Spectrometer. This method uses reagents mixed with the water sample to produce a coloured solution which is calibrated for the particular test (nitrate or phosphate). The light transmissivity through the sample is then measured to assess indirectly the chemical concentration.

Saturated hydraulic conductivity

The hydraulic conductivity is a measure of the ability of the soil to transmit water (Burt, 1978), a property which may be greatly affected by repeated drainage and cultivation. Saturated hydraulic conductivity (K_{sat}) is a fixed value indicative of the transmissivity of water in the soil under saturated conditions. The aim of incorporating K_{sat} investigations into the hydrological monitoring programme was to identify any differences either within or between wetlands, in their response to hydrological inundation, which may be attributed to anthropogenic disturbance.

This has important implications for the ability of wetlands to regenerate and to sustain agricultural production particularly when the height of the water table and the rate of movement of water within the soil are both limiting factors to plant growth. It was also envisaged that the assessment of K_{sat} values would aid the interpretation of indigenous drainage design and technology. As K_{sat} affects the optimum number and spacing of field drains (Landon, 1991), values could be compared to the farmers' perception and understanding of the drainage properties of the soil.

Of the several methods available for measuring K_{sat}, the auger hole method described by Van Beers (1963), was considered the most appropriate method for the purpose of the research. This method requires removal of the water from a dipwell column and then timing its recovery rate, i.e. the inflow of water into the tube as it reaches the height of the surrounding water table. The rate of recovery is then converted by the appropriate formula to a K_{sat} value that represents the average permeability of the soil layers between the top of the water table to the bottom of the dipwell.

This technique was carried out on all the study wetlands except those where the water level remained above or at the surface, making the removal of water from the dipwell column impossible. Tests were carried out for each dipwell within a wetland to enable the small-scale variation in K_{sat} to be identified as well as the overall site mean.

Analysis of rainfall data

The measurement and analysis of rainfall provides a means of estimating the amount of water entering the wetland hydrological system. Although the wetlands may receive input from groundwater, this is difficult to quantify and it is highly likely that the aquifer will be recharged by rainfall falling within the study area.

Five rainfall gauges are located in the study area and of these, one gauge is a component of a meteorological station operated by the Soil Conservation Research Project (SCRP) in the Dizi catchment, north of Metu, and two are run by the Ministry of Agriculture in Coffee Improvement Project (CIP) sites. The remaining two, at Gore and Metu, operate through the Ethiopian Meteorological Agency. Rainfall records from these stations were used to establish the normal pattern of rainfall throughout the year, enabling the representativeness of rainfall during the study period to be assessed. The impact of rainfall on the other hydrological parameters at different wetlands was also undertaken using the rainfall data from the nearest gauge.

Wetland surveys

An important part of the hydrological research programme was relating measurements of each parameter to patterns of drainage and the overall hydrological conditions of each wetland. It was, therefore, necessary to carry out basic field surveys of each site to establish the influence of topography and any other significant landmarks on the hydrological regime. The techniques used ranged from producing simple sketch maps based on observation during weekly site visits, to the use of a Global Positioning System (GPS) for accurately mapping the perimeter and drainage channels of each wetland. These were combined to produce comprehensive maps of each wetland.

Investigating indigenous wetland knowledge

One of the key aims of the research was to identify the indigenous wetland knowledge held by local communities, the ways in which this is operationalized and how this contributes to hydrological sustainability. The interaction of people and resources, which is characterized and driven by local knowledge, does not lend itself easily to a scientific analysis and can be more easily understood through a phenomenological research approach which centres on the meaning and significance individuals place on their wetland interactions. Participatory research methods arguably offer the means to pursue such an approach, given their effectiveness in eliciting qualitative contextual information, which cannot be measured easily by scientific methods.

The initial steps in formulating a methodology for investigating the indigenous wetland knowledge of farmers with respect to hydrological management, consisted of the identification of four main research areas in accordance with Objective 3 (see Chapter Four). The key questions surrounding the utilization of wetlands were summarized as:

1. What are the farmers' perceptions of the wetland hydrological system?
2. How do farmers interact with the wetland hydrological system?
3. How and why do farmers manage the wetland hydrology?
4. Where does wetland knowledge come from and how does it develop?

By using these as a framework for investigation, it was envisaged that the research could establish the characteristics of wetland knowledge, how it is operationalized in hydrological management practices and how this contributes to sustainable hydrological management. Question four also takes this further, addressing the question of the long term sustainability of hydrological management, where sustainability is rooted in the ability of wetland knowledge to adapt, evolve and cope in response to a range of pressures.

The use of Participatory Rural Appraisal (PRA)

Research methods in the arena of rural development have been constantly evolving since the 1960s when the majority of information was collected through the use of questionnaire surveys. This generated vast amounts of quantitative data which were analyzed using a variety of statistical techniques. One of the major changes to come about was the dawning realization that rural communities do not always conceptualize their lives within the limits of a questionnaire (Brace, 1995). Gradually, these institutionalized techniques of data collection came under increasing criticism (Brokensha et al., 1980; Chambers, 1983) and from this new participatory methods were developed. Chambers (1994a) lists the fields of participatory research, agroecosystem analysis, applied anthropology, field research on farming systems and rapid rural appraisal as precursors to what is regarded by many as the new panacea in rural development research, Participatory Rural Appraisal (PRA).

PRA evolved out of rapid rural appraisal (RRA), a technique which emerged in the late 1970s, itself having developed from several origins. The first was an increasing dissatisfaction with what has been termed 'rural development tourism' (Chambers, 1983). This phenomenon is characterized by brief rural visits by urban professionals which tend to be biased in terms of their spatiality, seasonality and gender which, according to Chambers (1993), means that rural poverty often remains unobserved. The second origin was the dissatisfaction with quantitative methods of data collection and finally, RRA developed from a need for more cost effective methods alongside the recognition that local communities have an exclusive body of knowledge (what became known as indigenous technical knowledge), which could be incorporated into development programmes.

The fundamental difference between RRA and PRA activities is that PRA empowers the community through increasing their level of involvement in any assessment of their own situation. It is by definition more participatory in nature rather than extractive and extends into the realms of analysis, planning and action. With this in mind, an often quoted definition of PRA has been:

> ...a family of approaches and methods to enable rural people to share, enhance and analyse their knowledge of life and conditions, to plan and to act (Chambers, 1994a, p. 953).

This aside, the key principles and methods of both RRA and PRA are similar. Chambers (1994b) suggests that they have the following in common:

1. there is a reversal of learning – researchers gain an insight into the local, physical, technical and social knowledge held by communities;
2. learning is a progressive and adaptive process – achieved through the flexibility of the approach;
3. biases are offset – especially those of rural development tourism. Both methods are relaxed not rushed, they involve listening rather than lecturing, and they include the poorer members of the community;
4. triangulation – which refers to the cross checking of information from different sources. This is achieved through applying a host of different techniques to the same research topic;
5. seeking diversity – which means looking for exceptions and investigating contradictions and anomalies. Variability rather than averages are sought.

There are, however, several differences which are mainly concerned with the behaviour and attitudes of outsiders in the PRA approach:

1. local people should be allowed to generate their own analysis and interpretation after the facilitator initiates the discussion;
2. self-critical awareness is important. Facilitators should examine their own behaviour and embrace criticism as an opportunity to learn;
3. practitioners of PRA should take personal responsibility for their activities rather than being accountable to manuals. The advantage here is that PRA can adapt to each unique situation;
4. the sharing of information is a fundamental issue, both at the community and international level.

To list the actual methods and mechanics of carrying out PRA would be beyond the scope of this book. Chambers (1997) provides a summary of the various methods available to researchers and a plethora of PRA manuals and guidelines exist as outputs of conferences, workshops and project experiences throughout the world (Chambers *et al.*, 1989; IIRR, 1996; Grenier, 1998). All the possible methods, however, tend to build upon the past experiences of PRA, especially those initiatives which have proved to be successful.

The use of diagrams and mapping techniques, in particular, are considered an integral part of PRA activities because studies have shown that local communities have a great capacity to map, model, rank and estimate features of their environment (Stewart, 1995). Because these are essentially visual phenomena, there is great opportunity to include all the members of a community in discussion and the outcome is an exchange of information between members of the community as well as with the researchers. These techniques can be taken one step further when the elements of time and historical change are introduced. For example, a map of forest cover could be developed and reconstructed to represent the changes occurring over a fifty year period.

Although having developed as a reaction against conventional research methods, PRA is by no means closed to the use of formal quantitative survey techniques. The approach regards the complementarity of techniques to be an important part of building up a more detailed picture of human ecology (Cornwall and Fleming, 1995; Scoones, 1995). Furthermore, by being flexible and using combinations of methods, a balance between academic rigour and the reality of individual situations can be reached (Abbot, 1997).

The main advantage of using PRA, especially in the study of natural resource management, is that it can accommodate the complex interrelationships between disciplines such as ecology, hydrology, economics and agriculture. It can be argued that only local people have a detailed, intimate knowledge of this complexity in their environment and because PRA operates at the local level, it constitutes a site specific and sensitive means of investigating and sharing this intimate knowledge. Whilst PRA does generate site specific information, it can, through its replication, draw attention to common elements rendering it a useful research tool in highlighting any generalizations. The result is a greater understanding of the social and environmental processes which affect the utilization of natural resources.

Participatory research remains a contentious issue. While it has been suggested that PRA provides the answer to many problems besetting conventional rural development research (Chambers, 1992), it has also been criticized for being biased and unreliable (Hammersley, 1992). Cornwall and Jewkes (1995) suggest that participatory research is rarely devolved completely into the community and in some cases the target community may be sceptical about investing their time and energy in PRA activities. Even if a programme of PRA is engaged, the enthusiasm of both local and outside participants may wane, especially if the preconceived ideas of the latter are not vindicated by research findings.

The over-adoption of RRA and PRA techniques has been another criticism. The view that PRA is an easy option requiring little training is a cause for concern, especially as the demand for training exceeds the supply of competent practitioners. It is often used as an easy, cost effective means of carrying out research, particularly when donor agencies demand its inclusion (Richards, 1995). The result is 'bad' PRA which, as it proliferates, becomes unrecognizable as a participatory, empowering and capacity building technique. Cornwall and Jewkes (1995) argue that such criticism does not devalue the role of participatory research.

It is they claim:

> ...ultimately about respecting and understanding the people with and for whom researchers work. It involves recognizing the rights of those whom research concerns, enabling people to set their own agendas for research and development and so giving them ownership over the process (Cornwall and Jewkes, 1995, p. 1674).

At the very least, PRA can be said to provide a much deeper insight into the lives of local people and for this reason, it was considered a critical research method for investigating indigenous wetland knowledge.

Operationalizing the participatory research

Having established the key research questions, these were operationalized into practical sessions designed to cover a specific research topic through the use of appropriate participatory methods. Numerous methods are available from the suite of PRA / RRA techniques widely used in recent research (Chambers, 1997; Binns *et al.*, 1997; Grenier, 1998) and each possess some degree of flexibility in terms of the circumstances under which they can be used but also the information they generate. Consequently, each technique can be applied selectively where conditions demand. In the following sections each component of the methodological framework is discussed alongside the participatory methods which were employed as a means of eliciting the relevant information. Running through the PRA programme, however, were several key themes relating to the implementation of the research.

First, focal subject sampling was used throughout the research. This technique was developed from social anthropology and involves the use of key informants, i.e. members of the community who have valuable information on the required topic of investigation. PRA was undertaken specifically with wetland farmers rather than a representative sample of the rural community as a means of providing relevant information on wetland hydrological processes and management techniques.

Secondly, familiarization was an important aspect of the research programme. In order to carry out PRA in a community successfully, it was important that an atmosphere of mutual respect and trust was created. Prior to the onset of the research, each community was visited and a discussion held about the aims and objectives of the research.

Thirdly, the triangulation of information was carried out to improve the robustness of the research findings. This was achieved as a result of the complementarity and overlap between the methods employed, which at times were able to elicit the same information and consequently cross check the validity of data.

Fourth, the semi-structured interview was prevalent. This technique, which is often regarded as the core of PRA, formed an integral part of the research as a means of clarifying and pursuing issues raised during specific sessions and activities.

Finally, and above all, participation played a critical role in the research proceedings. All those involved were given the opportunity to speak and lead discussions and activities. During each session, farmers were repeatedly kept up to date with the research and questions to the PRA facilitators were encouraged.

Farmers' perceptions of wetlands The first phase of the PRA research aimed to identify the relationship between indigenous wetland knowledge and the current scientific knowledge. It was logically assumed that hydrological management techniques would be a function of farmers' level of understanding and knowledge of wetland hydrology, hence it was important to determine the degree to which farmers understand the wetland in order to comprehend the management techniques and technology in operation. In the wider context, it was also necessary to understand the degree to which farmers act upon on their hydrological knowledge. Fundamentally, this information is the basis of indigenous wetland knowledge. It provides the building blocks of management strategies and probably any interaction the farmer has with the wetland environment.

As a starting point, the technique of resource mapping was employed as a means of establishing a rapport with the farmers and also identifying their perception of their own environment. This technique helped identify what the farmers considered to be their important resources. Resource maps were constructed on the ground using a variety of local materials including different leaves, twigs and fruit (Figure 5.4).

The principal method of data collection at this level was the use of seasonal diagrams. These are generally used to highlight the interrelationships of variables and the timing of activities, in this case, changes in the wetland hydrology over the year. This technique was used with groups of farmers no fewer than four in number. Using the example of wetland water table, farmers were asked to represent the monthly changes over the course of a year, using twigs or *cheffe* laid out on the ground. The graph produced would then be discussed and amendments made by the farmers. The method was repeated for other variables including rainfall and these were then sketched onto paper. Following the production of seasonal diagrams relating to the wetland hydrology, semi-structured discussions were carried out, focusing on the theme of seasonal hydrological variation. During these discussions farmers were encouraged to comment on any spatial hydrological variation they perceived within the wetland and to offer their views on the reasons for this variation.

The characteristics of farmer-wetland interaction The next level of the wetland knowledge research framework addressed the way in which farmers interact with the wetland system. The aim of this was to build up an inventory of the different practices and technology being utilized, particularly with reference to hydrological management. In addition, it aimed to establish the ways in which wetlands have been utilized in the past and determine how the technology and methods contrast with present day utilization. Furthermore, the characteristics of farmer organization and co-operation in wetland utilization were also explored.

Much of this part of the research was carried out with reference to the checklist of wetland utilization variables (Table 5.1) where the aim was to gather the relevant information available for each site. A great deal of this was collected through the observation and analysis of the wetland and the production of sketch maps, although several PRA techniques were used. The most appropriate was considered to be the seasonal calendar which was used to focus on the seasonal changes in specific wetland activities. This not only highlighted the range of utilization activities but also the methods and technologies involved. For example, taking drainage as the wetland utilization being studied, each activity in the drainage calendar could be represented graphically, followed by a discussion of the tools used for each activity and the labour requirements (Figure 5.5).

Another technique used was that of constructing time lines. These involve the identification through discussions, of significant events in the history of a community. The chronology produced is then used as the basis for discussing changes in a particular variable such as crop patterns in the wetland. The output is a historical documentary of the topic under study.

Strategies for wetland management Following the previous two approaches which described the characteristics of wetland knowledge, the research addressed the question of how this knowledge is operationalized in the form of wetland management strategies. Three key areas were addressed:

1. first, in order to understand the management strategies employed by farmers, the research aimed to identify the reasons behind the need to manage wetlands;
2. secondly, the research addressed the question of how specific management techniques and ideas are inherently linked and mutually supportive towards the fulfilment of the aims of the farmer, particularly when the wetland is used for a variety of purposes. The aim of this investigation was to understand how each farmer employs indigenous management tools / technology in combination, to produce a particular management strategy in accordance with their overall aims;
3. finally, the overall management strategies which operate in a wetland, which may affect the individual farmer's choice of utilization were investigated. This included the identification of organizational structures, community rules and established protocol in wetland management strategies.

The methods used for these investigations included transect walks and group discussions. Transect walks consisted of walking through each wetland with farmers, asking questions, observing and discussing a variety of features along the route. Although a provisional route was planned through the wetland, this was discussed before each walk and farmers were given the opportunity to change this depending upon what they considered important wetland features.

Figure 5.4 Farmers construct a resource map of Bake Chora wetland using a variety of natural materials

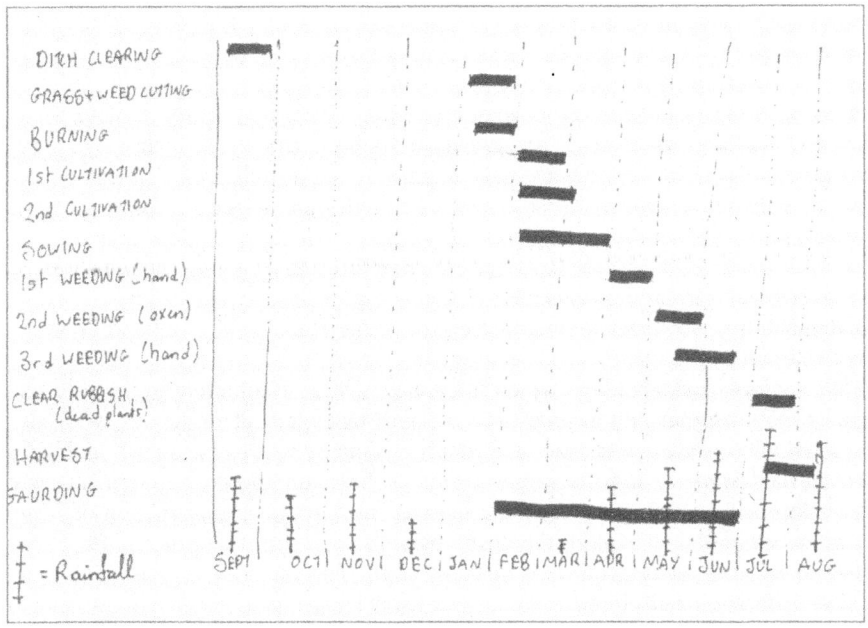

Figure 5.5 Field sketch of a seasonal calendar (including rainfall) produced by farmers at Dizi wetland

Table 5.1 Checklist of information collected during PRA sessions

Wetland use

Drainage	Cultivation	Grazing	Natural Uses

• drainage density
• drainage calendar
• channel characteristics
  - width
  - depth
  - length
  - layout
• (optimum) depth of drain below water table
• (optimum) water table depth for cultivation
• spacing of drains – minimum and maximum
• area under drainage & undrained
• maintenance techniques:
  - clearing vegetation
  - drain excavation
• structures for hydrological control
• water storage areas
• organization of drainage
• tools and equipment
	• crop selection
	• area under cultivation
	• crop maintenance
	• wetland access and ownership
	• rotation /intercropping techniques
	• timing of crop activities (for each crop type)
		• cattle ownership
		• duration / methods of grazing
		• vegetation preference
		• management of vegetation
		• tethering technology
		• rotational grazing
		• cycle of animals
			• water supply
			  - wells
			  - springs
			  - organization
			  - access
			• reed collection
			• cultural / religious significance
			• medicinal use of plant species

The acquisition of indigenous wetland knowledge This final level of enquiry into farmers' wetland knowledge attempted to identify the mechanisms by which indigenous wetland knowledge is acquired, organized, communicated and operationalized. It investigated the existence and significance of communication networks in which farmers exchange information on hydrological management practices, and in addition, the extent to which hydrological management techniques are generated, farmer experimentation being a potential source of change. More fundamentally, the research sought to establish the reasons why wetland practices need to evolve.

In addition to knowledge which is created within a community, the role and impact of 'external' knowledge on wetland management strategies was also explored. Those ideas and practices which originate from sources such as NGOs and the government, and the extent to which these are adopted by farmers were also addressed. A critical technique employed here was the construction of Venn diagrams designed to explore the changing influences on the farmers' use of the wetland and the sources of wetland knowledge. The method consists of farmers laying different sized circles (constructed from local materials) on the ground to represent the degree of influence of external actors on themselves. The farmer is represented by a circle in the middle and the other rings are laid adjacent to or intersect this middle circle This method can indicate the size or power of the external actor, as well as the degree of influence, generating a considerable amount of information.

Apart from Venn diagrams, semi-structured interviews and group discussions were carried out building upon the information collected from previous activities as a starting point for discussion. For example, where the seasonal use of tools were discussed in a previous session, the findings were presented back to the group of farmers, and a discussion was held around the subject of tool origin and design.

Limitations of PRA in the study wetlands

Participatory methods are not an infallible means of collecting information. At the simplest and most obvious level, they are prone to bias on the part of the facilitator and misunderstanding by both parties, especially where translators are used to convey questions, activities and responses. For this reason, the methodology relies on encouraging farmers to take control and present their knowledge in a graphical, easily understandable format wherever possible. These still present problems with interpretation. For example, seasonal farming calendars and seasonal diagrams of wetland hydrology presented problems with interpretation. In the latter, the major cause of contention is the question of how the farmers perceive rainfall: whether the graphs represent the volume of rainfall within each month or simply the number of rainfall days. In addition, although farmers were asked to illustrate the average annual pattern, this may have been misunderstood and instead, the previous year's rainfall may have been illustrated.

It was also difficult to establish if farmers were deliberately misleading the group facilitator or interpreter. Throughout the PRA programme, farmers from all the wetlands, relayed their concern over the nature of the research and they needed constant reassurances that ultimately the research would be of benefit to them. This is, of course, understandable, particularly in view of the treatment the farmers have received in the past from successive governments. Consequently, at the very least, the research was carried out in an air of suspicion on the part of farmers, although this tended to vary between sites. It must be recognized, therefore, that farmers did have an incentive to convey misleading and inaccurate information on the grounds that they were protecting their own livelihoods and in particular, their wetland resource. Clearest examples of this behaviour was a tendency to exaggerate farming problems, possibly in the hope that those involved in the research would be able to provide direct or indirect assistance.

Conversely, the self-respect of the farmers is an important consideration. On many occasions farmers were reluctant to acknowledge that they received any assistance from their fellow farmers or from outside agencies. Assistance from the MoA on farming techniques was often dismissed as inappropriate and useless and when questioned about the influences on their farming methods, many farmers stated that they had taught themselves. Similarly some farmers may have been too embarrassed to admit to practices and technology which may seem out of date to western researchers and even their fellow farmers. For example, the use of wetland plants for medicinal purposes was acknowledged by farmers at only one site. Equally, experimentation is a sensitive subject and often farmers may not be willing to draw attention to their efforts, especially those which failed.

In essence, there is no clear way of establishing whether farmers were being completely honest in most cases. In such circumstances, the only means of verification is the triangulation of information from other sources and the identification of trends in the information collected.

Summary

This chapter has outlined the methodological process of obtaining a representative sample of wetlands for intensive study, from the wider context of wetlands in Illubabor. Narrowing the range of variability was achieved by selecting a study area within Illubabor and those wetlands located within this area were classified to identify the key characteristics of those undergoing agricultural development.

The selection of sites under different levels of development together with the hydrological methods described, also provide the basis for establishing the impact of agricultural utilization on the hydrological regime of wetlands in the area. The results of this hydrological monitoring programme, which centre on the behaviour of wetland water table levels, are discussed in Chapter Seven.

In addition, the discussion of participatory research methods as a well established means of investigating the dynamics and characteristics of indigenous knowledge among rural communities, have also been discussed. A framework for investigating the role of IK in wetland management which incorporates these methods has been presented, the findings of which are discussed in Chapter Eight.

Chapter 6

The Study Wetlands

Introduction

This chapter presents the results of the selection of study wetlands, highlighting a typology of wetlands which was produced from the classification process described in the previous chapter. In addition, a model of wetland development is proposed on the basis of the occurrence of wetlands in Illubabor under a range of different land uses. The general hydrological and land use characteristics of the study wetlands are also presented using information elicited from local community members throughout the course of the PRA research programme. This is supplemented by information recorded during other field visits to the study area and by information compiled from other recent research projects, notably the Ethiopian Wetlands Research Programme (EWRP). Following this, the characteristics of each study wetland are discussed, providing a context for the discussion of the research findings in Chapters Eight and Nine.

The results of wetland classification in the study area

Wetland size

The wetlands in the study area were classified into three groups on the basis of their occurrence, these were: small (up to 0.29 km^2), medium (0.29 km^2 – 1 km^2) and large (above 1 km^2) (Figure 6.1).

Hydrological characteristics

The classification of hydrological characteristics revealed that those wetlands with both an inflow and outflow (termed mid-valley wetlands) were the most abundant in the study area (Figure 6.2).

Wetland order

Those wetlands with none preceding them in their catchment were found to be the most common in the study area (Figure 6.3). These were classified as 'headwater' wetlands.

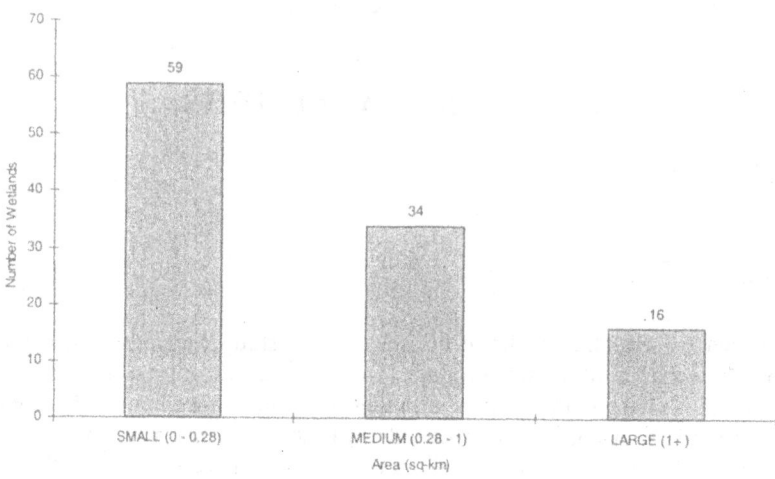

Figure 6.1 The final classification of wetlands into three size categories

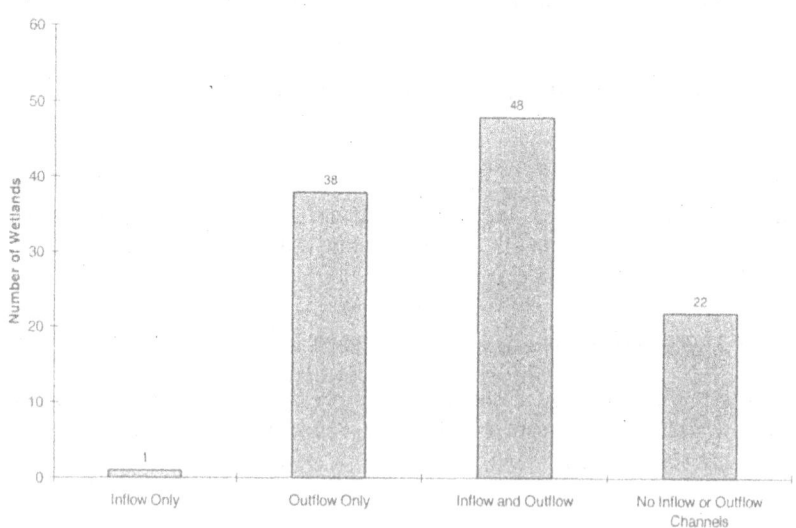

Figure 6.2 The inflow – outflow characteristics of wetlands in the study area

The Study Wetlands 97

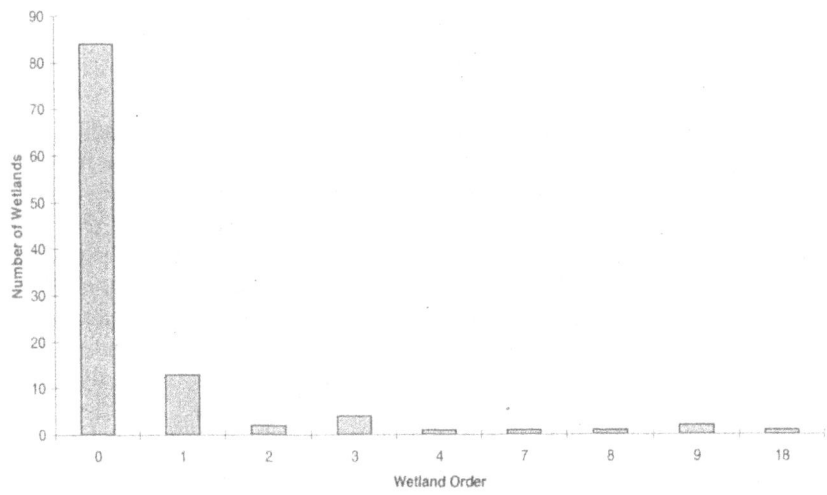

Figure 6.3   Wetland order in the study area

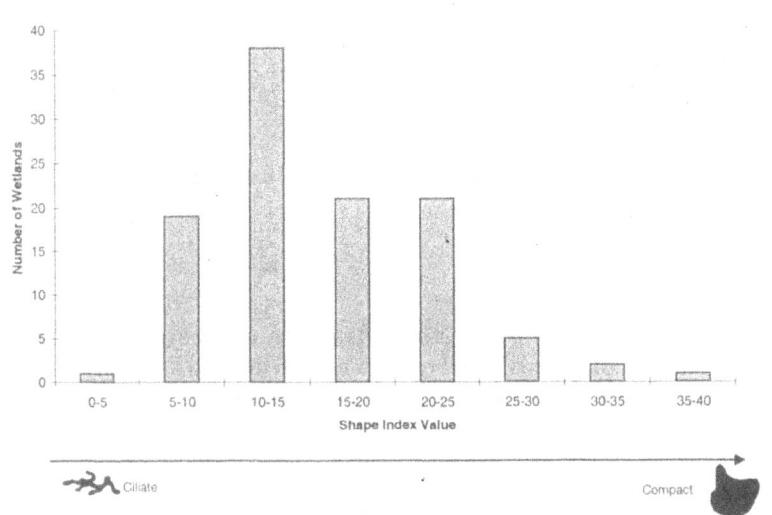

Figure 6.4   Wetland shape in the study area

Wetland shape

Most of the wetlands were found to have complex shapes rather than being rounded and compact, as would be expected in an area of predominantly steep sided, narrow valleys (Figure 6.4).

The wetland typology

Of those variables measured for each wetland the classification established that small wetlands, wetlands with none preceding them, complex shaped wetlands and wetlands with both an inflow and outflow were the most common in the study area.

The next stage of the classification sought to elaborate on the relationship between these variables with the aim of eliciting distinct *types* from the data. This involved analyzing the data by overlaying sets of information on the GIS and producing new coverages of the spatial distribution of wetlands where variables were linked. This method was repeated a number of times with combinations of variables until a typology of wetlands began to emerge. The resulting typology is shown below in Table 6.1

Table 6.1   The wetland typology results

	Wetland Type	Abundance	% of Total wetland area
1.	Small headwater	22%	6%
2.	Small mid-valley	18%	5%
3.	Medium mid-valley	17%	18%
4.	Small closed	16%	3%
5.	Medium headwater	7%	9%
6.	Medium closed	5%	6%
7.	Large headwater	5%	15%
8.	Large mid-valley	8%	33%
9.	Large closed	1%	2%
10	Unidentified	1%	3%

The results of the cluster analysis undertaken on the wetland data generally verified the existence of these wetland types, although the statistics suggested some degree of overlap in terms of wetlands belonging to distinct groups (see Dixon, 2000).

Ground truthing results

The ground truthing confirmed that 'small headwater' wetlands were by far the most abundant type of wetland in the study area and the most common type of wetland undergoing development activity. The ground truthing of the 'small closed' wetlands revealed that they all exhibited a small outflow stream channel, requiring a reclassification into 'headwater wetlands'. In terms of the initial classification, this greatly increased the number of 'small headwater' wetlands in the study area. 'Small mid-valley' wetlands were found to be the second most abundant type. Both these and 'small headwater' wetlands tended, however, to be under-represented by the 1:50 000 scale maps, presumably because of the misinterpretation of the aerial photographs on which the maps were based. Most of the narrow river corridors in the area were also found to contain small wetlands which were commonly under cultivation. On the basis of these observations, small headwater and small mid-valley were considered the most appropriate for further investigation.

Wetland development in the study area

Initial investigations into the ways in which wetlands are used were also carried out during the ground truthing. These suggested that the wetlands in the study area could be categorized into five stages of development (Figure 6.5).

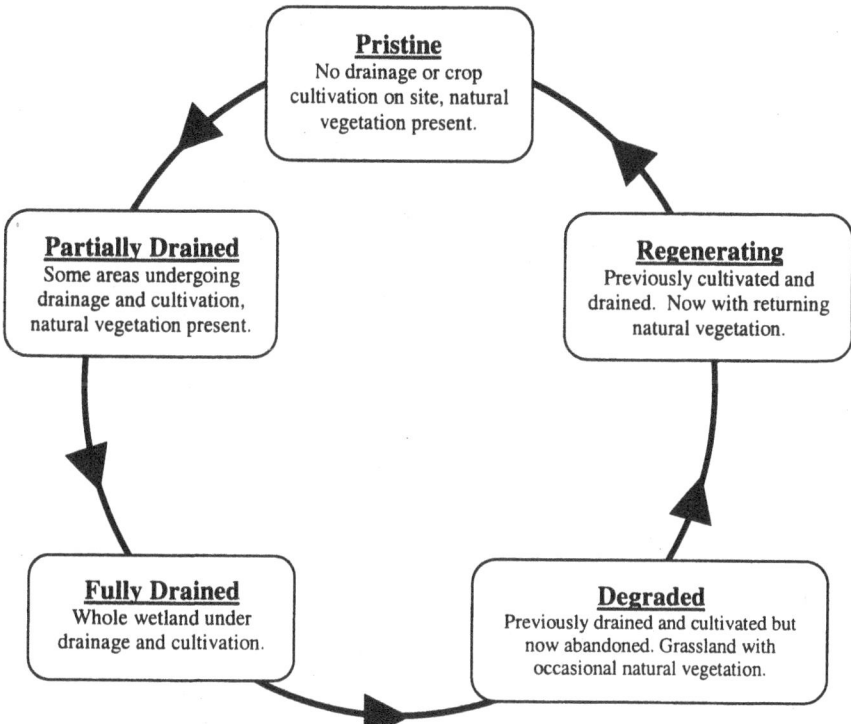

Figure 6.5 A conceptual model of wetland development based on the observation of wetland characteristics in the study area

The study wetlands

Small headwater wetlands were considered to be the most appropriate for further investigation on the grounds that they are the most common wetland type throughout the study area and also the most common type undergoing development. Consequently, five small headwater wetlands representing the five different levels of land use (Figure 6.5) were selected for hydrological monitoring, providing the main focal point of the research.

To enable comparisons to be drawn, two small mid-valley wetlands representative of a range of development stages were also selected (Table 6.2). Following the initiation of hydrological research activities on these wetlands sites, one further small headwater wetland (Supe) was incorporated into the PRA research programme, primarily as a result of other on-going research activities being carried out under the co-ordination of EWRP. This additional wetland was classified as partially cultivated. The location of all these sites within the study area are shown in Figure 6.6.

In reality, it was not possible to certify any of the sites in the study area as 'pristine', as there was no clear historical evidence that these wetlands had remained untouched by the human community. This class was, therefore, taken to represent those wetlands which exhibited the natural characteristics of a valley bottom wetland ecosystem. Those wetlands classed as 'regenerating' differ from 'pristine' sites in that the natural *Cyperus latifolius* vegetation may appear stunted, less abundant and less dense, whilst there may remain some indication of previous development activities. Although a 'degraded' wetland may ultimately regenerate, it is typically characterized by an environment which shows signs of alteration as a result of development. In particular, the presence of a different flora which may be adapted to drier conditions and the absence of *Cyperus latifolius* are key indicators of this development stage.

Table 6.2 Study wetlands and their development stages

	TYPE –1 (small headwater wetlands)	TYPE –2 (small mid-valley wetlands)
Pristine	Chebere	Anger (lower)
Partially cultivated	Wangeneye	Anger (lower)
Fully cultivated	Bake Chora	Dizi
Degraded	Hurumu	Anger (upper)
Regenerating	Tulube	Anger (upper)

General characteristics of selected wetlands in the study area

The study wetlands are located within Metu and Yayu-Hurumu *weredas* in the highland plateaux of Illubabor zone at elevations of between 1560 m and 1780 m

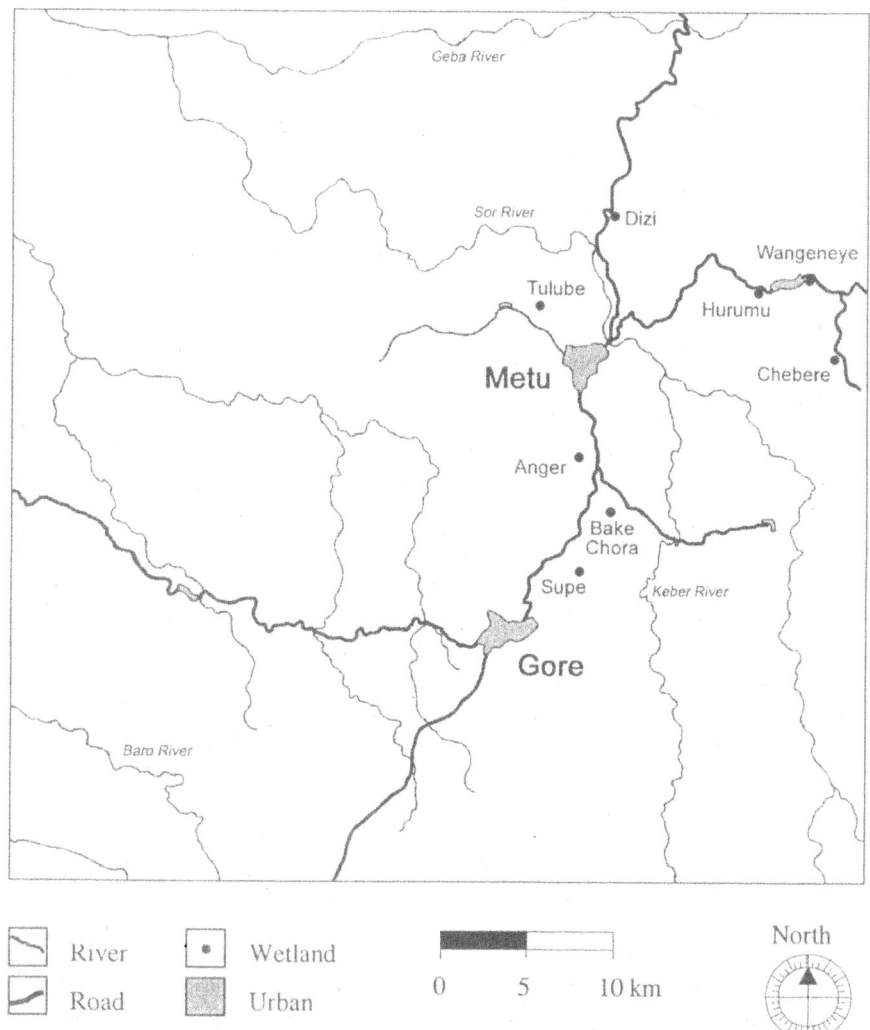

Figure 6.6   Location of the study wetlands within the study area

a.s.l. (Table 6.3). The wetlands are representative of the majority of those located within this region, typifying small, seasonally and permanently inundated areas located in valley bottom positions although usually at the valley head.

Table 6.3 The location and general characteristics of each study wetland

Wetland	Location (Astronomic)	Altitude (m asl)	Size (ha)	Average slope (°)	*Kebele*
Chebere	8°18'00"N, 35°43'40"E	1780	10	1.10	Gaba
Wangeneye	8°20'36"N, 35°42'49"E	1780	6	1.35	Wangeneye
Bake Chora	8°12'42"N, 35°36'00"E	1700	8	1.66	Bake Chora
Hurumu	8°20'19"N, 35°40'29"E	1760	4	1.92	Hurumu
Tulube	8°20'00"N, 35°33'00"E	1680	8	0.98	Tulube
Dizi	8°23'07"N, 35°36'24"E	1560	4	1.52	Dizi
Anger	8°15'00"N, 35°34'40"E	1640	16	1.12	Kawona Catu
Supe	8°12'00"N, 35°35'20"E	1720	10	1.53	Bake Chora

Hydrological characteristics of the study wetlands

Each of the study wetlands have similar characteristics in that they are classified as small (less than 0.29 km^2) and are located in the upper reaches of river systems being either headwater or mid-valley wetlands (Table 6.4). Using the hydrological classification proposed by Gilvear and McInnes (1994), all of these wetlands are characteristically *flushed seasonally omnitrophic* wetlands, assuming the likelihood of a groundwater influence in wetland recharge and discharge. Water inputs are from rainfall, surface water and groundwater, whereas outputs are from evapotranspiration, surface water and groundwater. The influence of surface water is, however, variable as headwater wetlands typically receive much of their input from springs which are directly influenced by rainfall, infiltration rates and sub-surface flow in their small catchments.

In contrast, mid-valley wetlands are located in the mid-reaches of a larger catchment and characteristically possess well-formed, natural stream channels that provide the main source of these wetlands' water supply (although according to most farmers, springs are also influential). According to Howard (1997) these wetlands are typical of the range of valley bottom wetlands found throughout East Africa and in terms of their ecohydrology, can be classified as permanent, spring fed / riverine, sedge swamps.

Soil characteristics of the study wetlands

Several studies have been carried out on the soil characteristics of some of the study wetlands (Yizelkal Fantahun, 1998; Solomon Tekalign, 1998; Asmamaw Legasse, 1998; Belay Tegegne 1998), namely Chebere, Tulube, Dizi, Hurumu and Wangeneye. The soils of these wetlands are typical of gleysols or soils with gleyic properties, i.e. they are influenced by flooding over long periods. Consequently, the soil properties reflect anaerobic conditions which facilitate the accumulation of

Table 6.4 The hydrological characteristics of each study wetland

	Hydrological Class.	Main water source	Drainage	Key hydrological conditions
Chebere	SH	Springs / runoff	no clear channel	high water table throughout year
Wangeneye	SH	Springs / runoff	artificial drainage	spatially variable water table
Bake Chora	SH	Springs / runoff	artificial drainage	low water table throughout year
Hurumu	SH	Springs / runoff	artificial drainage (abandoned)	Permanent and very low water table
Tulube	SH	Springs / runoff	intermittent channel	high water table throughout year
Dizi	SMV	Inflow stream	natural stream & artificial drainage	low water table
Anger	SMV	Inflow stream	natural stream & artificial drainage (some abandoned)	spatially variable water table
Supe	SH	Springs / runoff	artificial drainage	spatially variable water table

SH Small Headwater
SMV Small Mid-Valley

organic matter near the soil surface. The depth of the study wetlands' soils tend to range from a minimum of 60 cm to more than 200 cm. The soil profile is generally characterized by a darkly coloured A horizon which is rich in partially decomposed organic matter and the roots of plants.

The various studies highlighted above suggest that the texture of the soil ranges from sandy loam and sandy clay loam in the A horizon and from sandy clay to clay in the subsoils. The bulk density of soils within the soil profile have been reported as ranging from between 0.5 g cm^{-3} to 0.8 g cm^{-3} in the topsoils (0 – 80 cm), with a tendency to increase with depth up to 1.5 g cm^{-3} in the lower horizons (80 – 160 cm) (Belay Tegegne, 1998).

Utilization of the study wetlands

Having been selected to represent the range of land use characteristics, the study wetlands range from having a pristine status to those under full cultivation (Table 6.5). The former are characterized by an extensive coverage of natural vegetation (*Cyperus latifolius*) known locally as *cheffe*, which is harvested for a range of uses although principally for the construction of roofs. In addition, these pristine sites are often used during the dry season for the grazing of cattle which are given unlimited access.

By contrast, those wetlands which are undergoing cultivation are characterized by a network of artificial drainage channels. Such a network has the dual function of conducting water away from the springs located around the valley sides thereby reducing the residence time of water in the wetland, and also reducing the

Table 6.5 The land use characteristics of each study wetland

	Development Classification	Main Use
Chebere	Pristine	*cheffe* harvest / some grazing
Wangeneye	Part Cultivated	*cheffe* / maize crop
Bake Chora	Fully Cultivated	maize crop
Hurumu	Degraded	grazing
Tulube	Regenerating	grazing
Dizi	Part Cultivated	maize crop / tef
Anger	Multi - use	grazing / *cheffe* / maize
Supe	Part Cultivated	maize crop / sugar cane / *cheffe*

saturation of the wetland soil and consequently the height of the local groundwater table. In mid-valley wetlands, natural stream channels are commonly utilized as main drains into which artificially constructed lateral channels flow, whereas in headwater wetlands, a main central channel is excavated by farmers.

Drainage of the saturated wetland soils facilitates the cultivation of a variety of crops, although in the study wetlands maize dominates the wetland farming system. In some areas, however, sugar cane, tef and vegetables, including cabbage and potato, are also grown in small quantities from year to year. Depending on the variety of maize used, sowing takes place between January and March, and the green maize is usually harvested between June and July during the traditional hungry season (Figure 6.7). Following the harvest of maize, the residual vegetation is either collected and burnt or allowed to decompose. Providing the rains subside and the wetland soil moisture recedes, the residual soil moisture can facilitate the cultivation of tef in the wetlands. This is usually sown in late September and harvested at the end of November or early December. If tef is not cultivated during this period, the wetland may remain fallow and in some cases it is used as rough grazing until the period of maize cultivation begins again. Giving cattle access to these areas during this time is regarded by many as an effective means of increasing the nutrient input into the wetland soil as a result of dung production.

The system of wetland farming includes a variety of specific tasks that are undertaken throughout the calendar in order to ensure effective drainage and maximum yields. These farming practices vary from site to site but generally tend to involve periods of drainage channel maintenance, weeding, sowing, burning, ploughing, crop selection and throughout the whole process, pest guarding.

Those wetland sites classified as regenerating and degraded are by contrast, free of any direct agricultural intervention although this may have been influential in the past. A key characteristic of degraded wetlands is effectively their drier conditions which, together with their nutrient poor status, support a different range of vegetation. These conditions are, however, suitable for the grazing of cattle which represents the major land use in these sites.

Where the wetland shows signs of regeneration, there is typically a return of some *Cyperus latifolius* and often several intermediate weed species in conjunction with the partial restoration of natural hydrological conditions. Whilst these sites

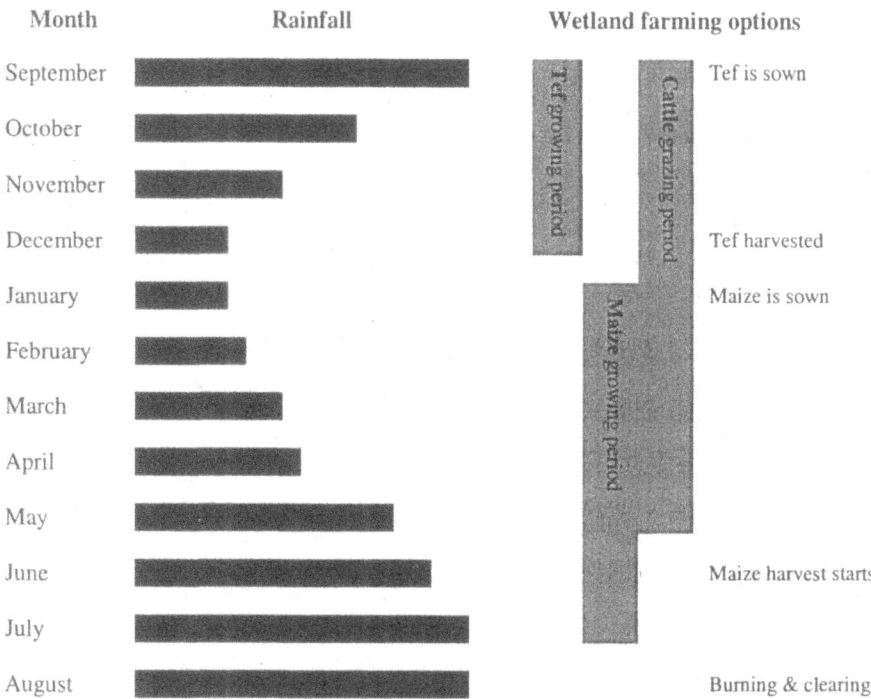

Figure 6.7 A wetland agricultural calendar showing the timing of the main land use options

may experience a return to periods of hydrological inundation which may ultimately emulate that of a natural wetland environment, the extent to which these sites do fully regain their pristine characteristics is, however, ambiguous and this is explored further in Chapter Seven.

Chebere wetland

Chebere wetland (Figure 6.8) is a pristine, headwater wetland and is located to the south-east of Hurumu along a dry season road leading to the small village of Baro (30 km east of Metu). It is located within the *kebeles* of Antona Chancho and Ordin in the *wereda* of Hurumu - Yayu. The wetland is characterized by dense *cheffe* vegetation which grows to a height of 2 metres in places, and severe waterlogging in the centre, although farmers suggest that the level of flooding is lower than in the past. No clear drainage channels are located within the wetland although there is an intermittent, seasonal stream at one side which drains to the south. Here the wetland drains into a stream which flows into the Sor River some 8 km away. According to farmers, the main sources of water in the wetland are the six springs which are located around the sides of the wetland.

The soil in Chebere wetland extends beyond a depth of 160 cm (Belay Tegegne, 1998). Soils at the top of the wetland tend to be sandy clay loam in texture with bulk densities ranging from 0.6 g cm^{-3} (0 – 30cm) to 1.18 g cm^{-3} (100 – 130 cm) down the soil profile. In contrast, soils towards the middle reaches of the wetland are dominated more by clay material in subsurface layers, although the topsoils are characteristically rich in organic material. Towards the outlet of the wetland where the *Cyperus latifolius* is less dense and abundant, the topsoils exhibit a relatively high bulk density (1.26 g cm^{-3}) in contrast to the preceding wetland areas (0.42 – 0.88 g cm^{-3}). This is similar down the soil profile in this area where bulk density increases to 1.44 g cm^{-3} and soil texture ranges from sandy clay to clay loam at a depth of 150 cm (Belay Tegegne, 1998).

Although initially classified as pristine, subsequent research suggested that this wetland had in fact been cultivated in the past and that drainage and cultivation of the wetland was initiated during the later reign of Haile Selassie. The major crop in the wetland in the past was maize although small quantities of vegetables including potato and cabbage were also grown. Several varieties of maize were used throughout the period of wetland cultivation, starting with *habasha*, a local variety which matures in six months, during the Haile Selassie era. This was followed during the Derg era by the cultivation of *jimma* and *bako* varieties which mature in five months. The most recent to be used is the *kenya* variety, which matures in four months and which is now widespread throughout the study area. Farmers suggest that crop productivity began to decline three years after cultivation commenced as a result of inadequate drainage and declining soil fertility.

The cultivation of Chebere wetland ceased in 1991 following the change in government and the community's decision to stop cultivating for reasons which included a need for both *cheffe* and new grazing areas for cattle. Although there remains a food shortage during the hungry season, farmers have since 1991 sold their labour in order to buy food during this period. During the 1997 – 1998 study period Chebere wetland continued to be used primarily as a source of *cheffe* for thatching and ceremonial purposes but also as a grazing area for small numbers of cattle. In addition, farmers also collect water from the wetland for their own use and several medicinal plants such as *gondi*, which is crushed and used for the treatment of headaches. In 1999 the head of Chebere wetland was drained by local farmers on the instructions of the Ministry of Agriculture. This was reportedly undertaken as part of a wider strategy to improve food security throughout Illubabor, in response to several years of unreliable rainfall and impending drought conditions (pers. comm. Patrick Abbot, 1999).

Wangeneye wetland

Wangeneye wetland (Figure 6.9) is a partially cultivated headwater wetland located approximately 20 km east of Metu, outside the village of Hurumu. It is situated on the northern edge of the larger Yobi wetland, within a small catchment which is dissected by the main Metu – Bedele road. The catchment is characterized by steep slopes down to the wetland and is composed of small areas of crop

Figure 6.8   Chebere wetland

Figure 6.9   Wangeneye wetland (under maize cultivation)

cultivation, coffee forest, eucalyptus, scrubland and grazing land.

At the head of the wetland is an uncultivated area of land, characterized by saturated conditions and dense *cheffe* growth. This is separated from the cultivated area by a deep latitudinal drainage channel which conducts water into a central main drainage channel. The relatively densely spaced system of drains in the wetland are never more than 1m in width or 75 cm in depth and in some places, rocky outcrops prevent the excavation of any channel. Farmers report that runoff from the nearby Metu – Bedele road represents a significant source of hydrological recharge and since this was built, drainage problems and sedimentation in the wetland have increased significantly. The main drain terminates at a point further down the wetland in an area of *cheffe* vegetation, although in 1999 this area and the headland *cheffe* were subsequently drained. The outflow of the wetland is a natural stream channel which flows into Yobi wetland.

Despite the occurrence of surface rock throughout the wetland, the soil depth ranges from approximately 226 cm to less than 60 cm. Texture ranges from sandy loam at the surface to sandy clay loam in the lower soil layers. Bulk density down the soil profile exhibits a narrower range of variation than at other measured sites, ranging from 0.54 g cm^{-3} ($0 - 37$ cm depth) to 1.09 g cm^{-3} ($63 - 82$ cm depth) (Belay Tegegne, 1998).

The wetland has been under full cultivation since the reign of *Dejazmach* Tassaw Walallu, the governor of Illubabor between 1945 – 1957 who, according to local farmers, ordered the cultivation of wetlands in response to food shortages caused by the widespread investment in coffee and subsequently the displacement of traditional cropland. Since then, cultivation has continued to the present, although it was disrupted for two years during the Derg regime, when the villagization programme took place. According to farmers, the head of the wetland which is dominated by dense cheffe growth, was abandoned when the government introduced a regulation that the top 50 metres of a wetland should be conserved. They also claim that this area suffered from frequent flooding as a result of the concentrated runoff from the nearby road. During the 1997 / 1998 field season, four farmers were cultivating the wetland with maize, while others had abandoned their plots as a result of labour shortages which prevented them from draining.

Bake Chora

Bake Chora wetland (Figure 6.10) is located approximately 7 km south of Metu near the village of Kemisse, within the *kebele* of Bake Chora. The wetland can be classified as a small, headwater wetland which, during the 1997 – 1998 season, was fully drained and cultivated although several farming plots within the wetland were temporarily abandoned. The small catchment is composed of farmland (approximately 60 per cent), in which maize constitutes the major crop under cultivation, and forest (40 per cent) in which coffee is also cultivated.

Its utilization for crop cultivation means that the wetland is characterized by a network of drainage channels. A resource map produced by Bake Chora's farmers clearly illustrates the main drainage channels and the ownership of plots within the wetland, which are usually separated by smaller drains. These channels drain water from several headwater springs the largest of which is continuously maintained and excavated in order to supply potable water for local farmers. This main spring is also used periodically for washing.

The excavation of drainage channels, which are usually no more than 70 cm in width and 80 cm in depth, is carried out annually and their location tends to change from year to year depending on the intended land use. The position of the central main drain, however, remains the same each year, although this is maintained annually in response to sedimentation, collapse and weed infestation.

Located within a small limb at the head of the wetland, is an area of *cheffe* which is usually waterlogged and drained by a small channel. According to farmers, this and other small areas of *cheffe* which are a result of abandoned wetland plots, are used in the construction of *godo*: small huts which provide shelter when pest guarding takes place. Farmers maintain that the amount of *cheffe* provided by this wetland is insufficient to be used in the roofs of *tukuls*, the main source being a wetland known as Didu which is reserved specifically for *cheffe* growth.

The agricultural utilization of Bake Chora dates back to the period of Menelik II (1889 – 1913) when *Dejazmach* Ganame, then the governor of Illubabor, notified farmers that they should engage in wetland cultivation in response to a state of famine. According to farmers, their ancestors discussed this proposal then reached an agreement on how to drain the wetland and how wetland plots should be allocated. A local leader known as Sheikh Abdella is credited as the first person to introduce techniques and practices of wetland drainage to the farmers at this time (Tegegne Sishaw, 1998). Farmers maintain that the wetland has been cultivated each year since this period, even though they have noticed a steady decline in crop productivity.

During the 1997 – 1998 season wetland use is dominated by the system of maize cultivation, although in some plots at the head of the wetland, tef is grown after the wet season. Cattle grazing also take place on the wetland at different times throughout the year at the discretion of individual farmers, although during the dry season (December to March) the cattle benefit by grazing on the palatable grasses and sedges in the wetland.

Hurumu wetland

Hurumu wetland (Figure 6.11) is a degraded headwater wetland located approximately 18 km east of Metu, on the outskirts of the village of Hurumu. Formerly a demonstration site for MFM (1990 – 1996), the wetland is surrounded by a steep sided catchment which features extensive terracing cultivated with bananas, coffee and eucalyptus, which were initially established in 1990. Springs are the main source of water in the wetland, the largest of which is located at the head of the wetland from which a drain conducts the water through the wetland to the outflow. Vegetation in the wetland is typical of a dry grassland environment, although some areas adjacent to collapsed drains are waterlogged for much of the

110 *Indigenous Management of Wetlands*

Figure 6.10   Bake Chora wetland

Figure 6.11   Hurumu wetland

year and show signs of *cheffe* recolonization.

Soils in Hurumu are typically of sandy clay loam and sandy loam texture although at the head of the wetland, clay is the dominant material between a depth of 20 – 80 cm. The bulk density of Hurumu's soils range from 0.75g cm^{-3} in the upper layers (0 – 20 cm) increasing to 1.16 g cm^{-3} at depths of over 100 cm (Belay Tegegne, 1998).

Most of the wetland is covered with the remains of a 'herringbone' drainage system and a series of raised beds which were formerly used for the cultivation of a variety of vegetables. The drainage channels range in width from 50 cm to just over four metres in the central area, although this appears to have been a result of streambank erosion following the wetland's abandonment. Below the area still under the influence of the old drainage network, *cheffe* has started to recolonize and during the dry season of 1998 there was some evidence to suggest that this was being harvested by local farmers. The drainage channel here is intermittent until further downstream where, at the outflow of the wetland, it flows over a rocky outcrop prior to terminating in a waterfall which drains to the south-west through a narrow valley and eventually into a larger wetland.

According to MFM, this wetland was cultivated during the Haile Selassie era but abandoned at some later stage until MFM initiated its development as part of an agro-ecology programme in 1990. Under MFM's management, the site was used to demonstrate wetland agriculture which included the cultivation of vegetables and in one area, an attempt at fish farming. This continued until 1995 when their staff allegedly noticed that the wetland was beginning to dry up and the vegetable plots were being irrigated by hand. Cultivation of the site was subsequently abandoned and ownership was handed over to the nearby school who now use it as an 'ecological learning' area although cattle are allowed access to this area. At the outflow of the wetland which is not owned by the school, the stream channel is used for washing and bathing.

Since November 1999, subsequent research visits carried out by EWRP staff have revealed that the lower area of Hurumu wetland towards the outflow has been fenced off by local farmers who have denied their cattle access to this area. As a result, *cheffe* is allegedly recolonizing this area at an unprecedented rate (pers. comm. Patrick Abbot, 2000).

Tulube wetland

Tulube wetland (Figure 6.12) is situated approximately 6 km north-west of Metu, in an area of Tulube *kebele* known as Mendido Koya. This spring fed, headwater wetland has an intermittent stream channel which drains in a northerly direction to an outflow stream, which eventually drains into the Sor river. The small catchment is composed mainly of farmland with the remains of natural forest restricted to some locations along the edges of the valley bottom. The catchment also contains areas of forest in which coffee is cultivated and there is a small plantation of eucalyptus on the valley side.

Several springs are located around the valley sides, the largest of which is located at the head of the wetland. Farmers have excavated a drainage channel from this spring, which runs for several metres before dispersing into the dominant *Cyperus latifolius* vegetation. This vegetation varies in density and height, with some areas toward the bottom and centre of the wetland being over 1.5 metres in height. These areas are also characterized by waterlogged conditions which prevail throughout the dry season when other areas become much drier.

The depth of the soil appears to be shallow in the mid reaches of the wetland where in some places, rock protrudes from surface. According to farmers, the depth of the soil to the bedrock gradually became much shallower towards the end of the years in which they cultivated the wetland. Soil texture generally ranges from loam in the upper layers (0 – 28 cm) to sandy clay loam and clay loam with decreasing depth (up to 120 cm). Bulk density in the topsoils (0 – 20 cm) ranges between 0.75 and 0.93 g cm^{-3}, and this tends to decrease slightly (between 0.68 and 0.88 g cm^{-3}) at depths of 20 – 55 cm. At greater depths (up to 120 cm) bulk density increases again up to 1.16 g cm^{-3} (Belay Tegegne, 1998).

According to its farmers, Tulube wetland was first cultivated during the later years of Haile Selassie (1960 – 1974) on the instructions of the government in order to address the problem of food shortages. Full drainage and cultivation of maize continued throughout the Derg regime when wetland production exceeded their subsistence capacity and crops were sold at the market. Towards the end of the Derg regime however, farmers first began to notice a decline in crop productivity from the wetland. This and a host of other problems, in particular labour and oxen shortages, caused the wetland to be abandoned during the late 1980s. Following abandonment, *cheffe* began to recolonize the wetland to the extent shown in Figure 6.12 which represents the state of the wetland at the beginning of this research (1996).

Tulube wetland has been used primarily as a source of *cheffe* for the roofs of huts, granaries and bee hives since cultivation ceased. It is also utilized on several occasions throughout the year for ceremonial purposes. Figure 6.13 however, shows Tulube wetland in May 1999, clearly in a state of degradation in terms of its vegetative cover compared with Figure 6.12. Afework Hailu and Abbot (1999) report that Tulube's farmers at this time regard the supply of *cheffe* from the wetland as inadequate. Apart from the wetland showing an overall decline in *cheffe* growth, the remaining *cheffe* is also considered to be inferior in that it is short, soft and as a result it only lasts as roofing material for one year. According to farmers, better *cheffe* is available from the nearby Wuchi wetland. The main cause of this degradation according to farmers, is an increase in wetland grazing as a result of the loss of grazing resources upslope, although this situation will have been exacerbated by the effects of drought conditions on the wetland soil caused by the late onset of rains during 1998 and 1999.

Despite its critical use as a grazing resource, farmers prioritize its utility as a source of water, maintaining that the only alternative was the Sor River some distance away. As a result, the spring at the head of the wetland is protected by a fence and constantly maintained to provide a potable water supply. One further use of the wetland mentioned by farmers is the collection of medicinal plant species, which are used for treatments ranging from skin disease to toothache.

Figure 6.12   Tulube wetland (1996)

Figure 6.13   Tulube wetland (1999)

Dizi wetland

Dizi wetland (Figure 6.14) is located approximately 12 km north of the town of Metu, in the northern part of Dizi *kebele* situated adjacent to the main Metu – Alge road (Figure 6.6). The wetland is classified as a mid-valley wetland and is dissected by a well defined stream channel whose source is located approximately 3 km upstream. This central stream channel meanders through the wetland before becoming artificially channelled at the outflow of the wetland, where it flows under the main road and through a hydrological gauging station, which forms part of the Soil Conservation Research Project (SCRP) activities in the area. This stream eventually joins the Sor River some 2 km to the west.

In addition to farmland and isolated areas of forest, a resource map produced by Dizi's farmers shows that the catchment also contains several additional areas of wetland, the largest of which is known locally as Yember. Farmers also highlighted the presence of specific grazing areas within the catchment, as well as several areas of forest reserved for coffee cultivation. A resource map of Dizi wetland itself also revealed the presence of a government fruit cultivation farm located on the eastern valley side of the wetland.

In terms of hydrology, the primary source of water in the wetland is the central stream channel which, according to farmers, originates from two springs upstream. The wetland is also dissected by several man-made drainage channels in a layout which may be modified annually depending upon aims of the farmers. Despite these attempts at drainage, farmers report that drainage is inefficient in some parts of the wetland, particularly around the southern periphery. Farmers associate these problems with the presence of rocks which lie close to the surface in many areas and which impede the flow of water and present problems of waterlogging. In addition, the topographical variation inherent in the wetland surface and the general gradient of the wetland may influence the direction of runoff.

A detailed examination of the soil characteristics of Dizi wetland are provided by Yizelkal Fantahun (1998). In summary, the wetland soil can be classified as being of sandy clay loam in texture with the proportion of sand and silt decreasing down the soil profile. Bulk density initially decreases from 0.975 g cm^{-3} in the topsoil (0 – 16cm) to 0.65 g cm^{-3} in the sub-surface horizon (16 – 41 cm), then increasing with depth to 1.09 g cm^{-3} (up to 100cm). Acidity of the soil increases down the soil profile, with some areas reaching levels of pH 3.00.

Dizi is one of a number of wetlands in the *kebele* where drainage and cultivation was initiated during the 1950s, in order to secure food production during the 'hungry season' when there is a food shortage. According to farmers, wetland cultivation continued throughout the reign of Haile Selassie until the Derg period, when in 1976, wetland cultivation was scaled down in many areas because of labour shortages caused by military recruitment campaigns and farmers being required to work on communal farms. In 1991 widespread wetland cultivation was resumed following the downfall of the Derg government.

Discussions with farmers revealed a number of ways in which Dizi wetland has been utilized over the years. These have primarily included *cheffe* harvesting, water collection and crop cultivation. During the 1997 – 1998 study period the wetland was undergoing cultivation more intensively than at any other time in its history, even though the crop productivity of the wetland was reported to be at an all time low.

While maize and sugar cane constitute the major crops, maize is also intercropped with cabbage between January and June / July. After the maize harvest and as the rains recede, other vegetables are sometimes grown and in other wetlands within the catchment tef is cultivated during this period. In addition to cultivation, the wetland is also used as a water supply for washing (this was observed on several occasions). The farmers collect *cheffe* for their *tukuls* from Yember wetland located approximately 2 km to the south. Although also cultivated in the past, Yember wetland was abandoned when drainage became too difficult, again as a result of labour shortages but also because the construction of the Metu - Alge road at its outflow seriously impeded drainage. It now represents a major source of *cheffe* in the *kebele*.

Anger wetland

Anger wetland (Figure 6.15) is a small mid-valley wetland located approximately 6 km south of Metu, forming part of the Anger River catchment. The source of the Anger is located upstream in Supe wetland (Section 6.10) and it drains in a north-westerly direction to the Sor River. The catchment covers a large area of approximately 11.3 km^2, whose range of land use includes those typical of Illubabor as discussed in Section 4.2.3. The major land use in the immediate upland area surrounding Anger wetland is coffee cultivation which according to farmers' own estimates, accounts for nearly 30 per cent of the land cover.

The wetland itself is deeply incised by a meandering, well developed stream channel which reaches a width of three metres in some places, while becoming intermittent in areas characterized by dense *cheffe* vegetation. This channel, along with several springs located around the valley sides, provide the main source of hydrological recharge in the wetland.

Less information is available on the soil characteristics of Anger wetland, although Belay Tegegne (1999) indicates that the depth of the soils are in excess of 200 cm. The pH of the soils range from 3.96 to 5.64 although these vary throughout the site and at different soil depths. Texture ranges from clay loam in the topsoils (0 – 52 cm) with clay dominating the lower horizons.

Anger wetland during the 1997 – 1998 season was in a state of multiple use. Some areas were in a 'pristine' state and others ranged from abandoned land used for grazing, to plots of maize cultivation. There are conflicting accounts of the starting dates of cultivation although there is general agreement that wetland cultivation began in response to famine caused by drought. One farmer who claimed to be 106 years old, described how the wetland was first ploughed in 1949 after the end of the Italian occupation. Prior to cultivation however, the wetland was covered entirely with *cheffe* and there was no system of land allocation.

Following the Italian occupation, slaves from the south who had been working for the landlords were given their freedom and because they owned no land, they

Figure 6.14 Dizi wetland

Figure 6.15 Anger wetland

were allocated plots within the wetland. The wetland was then intensively ploughed until 1973 when farmers' attention turned towards an increasing shortage of *cheffe* and the wetland was mostly abandoned. At present, *cheffe* is prized more highly than the wetland's agricultural potential because of a general shortage of *cheffe* in the area.

Supe wetland

Supe wetland (Figure 6.16) is located on the southern tip of Bake Chora *kebele*, approximately 5 km south of Bake Chora wetland itself. The wetland is a partially cultivated headwater wetland, with some areas of *cheffe* located on the western margins and towards the outflow, which eventually forms the Anger River. As with the other headwater wetlands, the majority of the wetland's water supply originates in the various springs which are found along the valley sides. Drainage of the upper wetland area consists of a central channel originating from the main spring, which is joined by several lateral channels also originating from springs. This channel extends to the middle reaches of the wetland where the wetland becomes narrow and the soil is shallow. This area exhibits conditions of waterlogging which the farmers associate with the shallow soil and the occurrence of rock outcrops. This area was cultivated until 1997 when farmers abandoned it as a result of problems associated with flooding. Below this area, the wetland becomes much broader and is characterized by the remains of a drainage network and regenerating cheffe vegetation, the area having been cultivated with maize as recently as 1997. This area and the fringes of the upper cultivated wetland area, which are dominated by grasses, are now used as communal grazing areas.

Soils in Supe wetland range in texture from silty clay in the topsoils (0 – 16 cm) with a dominance of clay below this depth (up to 180 cm) throughout the wetland. The pH of the soils varies spatially and with soil depth, although the overall pattern is one of decreasing pH with depth (between 4.31 to 4.91 in the topsoils decreasing to 3.42 at a depth of 102 – 160 cm).

Drainage and cultivation were initiated on the instructions of the landlord some 70 years ago, during the Haile Selassie era. According to farmers, the landlord first became interested in a wetland farming demonstration site set up by a Koran teacher who was living in the area. He then ordered his tenant farmers to adopt this new technology on their wetlands and at the same time he gave them ownership of half their wetland as motivation for guarding against pests. If a farmer did not plough his allocated wetland area, it was given to someone else. The present day allocation of wetland plots dates back to the Derg government when farmers say, only the wealthier households were allotted a wetland plot. Consequently, many of the farmers in the Supe catchment are not wetland owners.

The present day utilization of Supe includes the cultivation of maize, sugar cane, beans and vegetables. *Cheffe* is also harvested from the areas marginal to those under cultivation and during the dry season, cattle are given access to these areas. In addition, springwater is collected for human consumption. During the 1997 – 1998 season, the fertility of the wetland is perceived as poor, which is also

Figure 6.16   Supe wetland

indicated by the presence of the plants *komate* and *tufo guracha*. As a result, crop yields are also poor, a problem which appears to be exacerbated further by a shortage of oxen, crop damage by wild pests and a lack of co-ordination between the farmers who own wetland plots. Furthermore, the need for *cheffe* in the area is quite high despite several wetlands nearby being reserved for this function. Supe's farmers, however, claim that while crop cultivation can be carried out on the uplands, *cheffe* is only ever found in wetlands.

Conclusions

This chapter began with a presentation of the variation in wetland characteristics found throughout the study area. From the wetland classification process described in Chapter Five several types of wetland were identified and selected for further study on the basis that they represented the dominant wetland characteristics in the study area. In terms of physical characteristics, small headwater and small mid-valley wetlands were found to be the most abundant. The majority of these wetlands in the study area were also found to show signs of anthropogenic disturbance, which was classified as either full cultivation, part cultivation, degraded or regenerating. In addition, pristine wetlands characterized by natural vegetation were also identified. The study wetlands selected and described in this chapter represent small headwater and small mid-valley wetlands under these development stages, thereby facilitating a representative sample of wetland characteristics for intensive study.

From the descriptions of the study wetlands, several initial conclusions can be drawn. First, it is clear that the practice of wetland drainage and cultivation dates back further than was initially suggested. Most wetlands appear to have been under cultivation during the Haile Selassie period and some for over 70 years.

Secondly, rather than declining crop yields being the major reason for the cessation of wetland cultivation, social and political factors appear to be more influential in determining wetland land use. The exception is Hurumu, where changes in the wetland hydrology prevented the continuation of crop cultivation.

Finally, there is some indication that the wetlands themselves are robust in that regeneration has taken place on some sites and the wetland initially identified as 'pristine' (Chebere) has regained its natural vegetative characteristics within a relatively short time period. Whilst vegetation can be used as an indicator of wetland regeneration, it is important, however, to establish the extent to which drainage and cultivation influence natural hydrological conditions and whether hydrological regeneration can occur. This issue is explored further in the following chapter where the findings of the hydrological monitoring programme outlined are presented.

Chapter 7

The Hydrology of Valley Bottom Wetlands

Introduction

This chapter presents and discusses the hydrology of valley bottom wetlands in the study area, based on one year's hydrological monitoring between August 1997 and July 1998. It starts with an analysis of the spatial and temporal characteristics of rainfall over the study area, assessing their implications for other hydrological variables. Following this, the general trends in the water table in each of the study wetlands is examined in relation to each other. This analysis of the water table data identifies a number of differences between wetlands and also differences within wetlands. The results of the hydraulic conductivity investigations are also presented and interpreted within the context of land use within each wetland. Following a discussion of the results of the hydrochemical monitoring programme, the chapter ends by addressing the impact of agricultural utilization on the hydrology of the wetlands.

Rainfall in the study area

Daily data were obtained from three rainfall gauges located relatively close to the study wetlands enabling estimates of daily and weekly rainfall over the study area to be made for most of the duration of the hydrological monitoring programme. In addition, a longer series of monthly records were obtained from these three and two other gauges, providing some indication of the longer term climatic conditions in the area.

Figure 7.1 indicates the average monthly rainfall between 1967 and 1997 at Metu and 1908 and 1998 at Gore, illustrating the unimodal seasonal pattern of rainfall which is characteristic to this region of Ethiopia. A comparison between this average pattern of rainfall and rainfall during the study period (using data from Dizi, Sor and Gore for the 1997 – 1998 season) is shown in Figure 7.2 and a comparison between each of these gauges is made in Figure 7.3. These figures show first, that rainfall during the study period is quite different from the long-term average, primarily in terms of the amount of rainfall occurring during each month. Mean annual rainfall during the 1997 - 1998 season (using data from Dizi, Sor and Gore) was 2012 mm in contrast to the 31 year average of 1925 mm as calculated from the average of the Dizi, Sor, Metu and Gore gauges during this period.

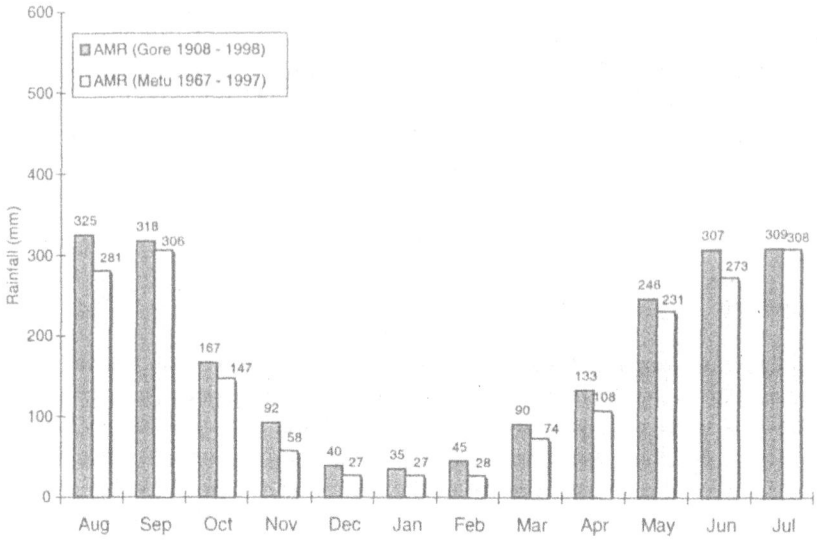

Figure 7.1   The average monthly rainfall in the study area (central Illubabor)

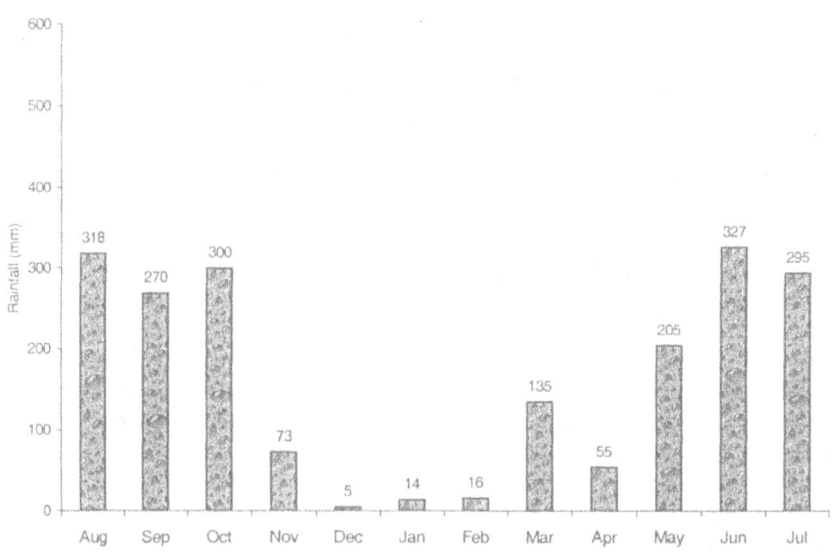

Figure 7.2   The average monthly rainfall (Dizi, Sor and Gore) during the study period (August 1997 – July 1998)

The Hydrology of Valley Bottom Wetlands 123

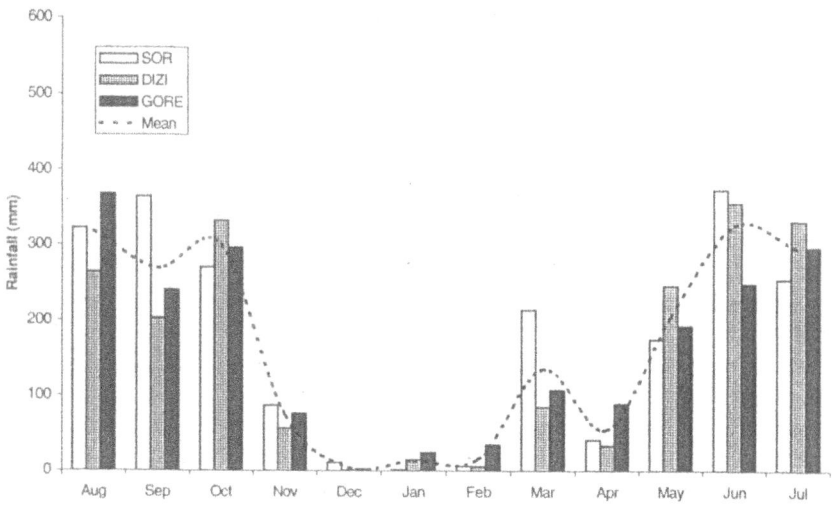

Figure 7.3 Monthly rainfall during the study period recorded at each gauge

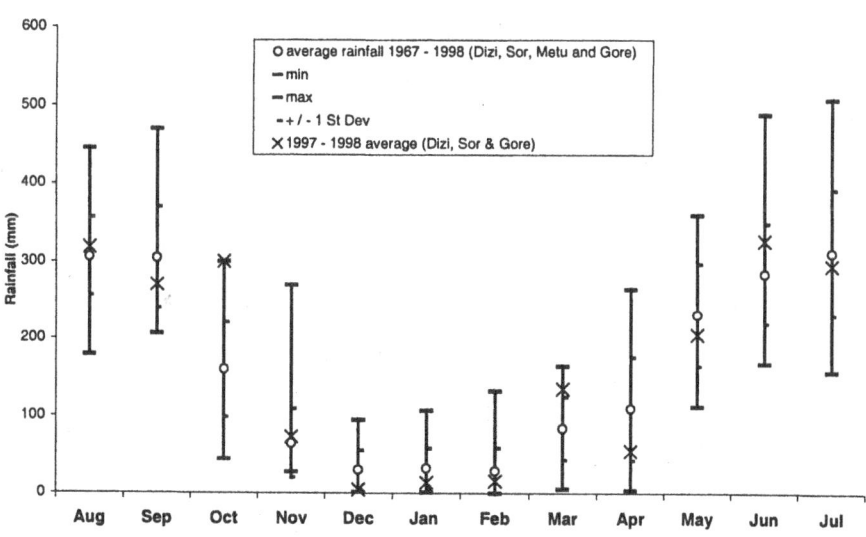

Figure 7.4 Rainfall during the study period compared to the 31 year average of Metu, Dizi, Gore and Sor

Compared with the descriptive statistics for the average of Dizi, Sor, Metu and Gore rainfall during the previous 31 years (Figure 7.4) it is clear that rainfall during the 1997 – 1998 season was characterized by unusually high rainfall during October, March and June. In contrast, April experienced well below average rainfall, with December, January and February also receiving slightly less rainfall than usual.

According to Jackson (1977) rainfall in tropical montane environments can exhibit extreme spatial variability with most rainfall occurring as localized rainstorms which vary in intensity and amount. During fieldwork this extreme spatial variability was observed on many occasions where for example, the town of Metu experienced a storm whereas the nearby sites of Tulube or Dizi received no rainfall.

Ideally, rainfall would be measured using a gauge at each wetland site, although in this research it was not considered practical given the problems encountered with maintenance and disturbance of other on-site hydrological equipment. Instead, using the existing rain gauges in the study area, the aim was to estimate rainfall at each study wetland using the nearest gauge. The narrow range of available data, however, meant that for comparisons with other hydrological data, only the average of relevant data at all the gauges at a particular time was used as a crude estimate of rainfall within the study area during the study period. For comparisons with farmers' perceptions of normal seasonal rainfall during the indigenous knowledge investigations, historical data from the nearest gauges were used.

Water table elevation in the study wetlands

The water table data

Hydrological data relating to water table levels were collected weekly using dipwells located in all the study sites between August 1997 and July 1998, representing one complete hydrological year. Within each site the number of dipwells contributing to the overall mean water table height varied each week, owing to theft, breakage and silting up of dipwell tubes. This tended to be much more pronounced on those sites where human interaction with the wetland occurred on a daily basis and where farming activities interfered with the equipment.

Where possible, missing dipwells were replaced immediately or sometimes during the following week and each replacement was recorded. The main implications of missing dipwells for the interpretation of data are primarily that the water table level expressed as a mean value may fluctuate more than would otherwise occur. The degree of fluctuation also depends upon the location within the wetland represented by the missing dipwells, e.g. dipwells removed from a wet location will induce a reduction in the mean water table level for the whole wetland and vice versa.

Seasonal variation in water table elevation – a composite view

The overall mean weekly measurements from all of the dipwells in all of the wetlands have been combined to show the seasonal pattern of water table elevation (Figure 7.5). In general, water table elevation was found to decline steadily from the wettest period in July and August, through to April when there is a sharp decline in water table elevation to almost 50 cm below the surface. With increasing rainfall in May, the water table elevation increased dramatically within a period of one month. At the end of the study period, during July 1998, the water table does not fully recover to the level of the previous July at which it was recorded one complete season earlier, despite the higher than average rainfall which occurred during June (Figure 7.4).

The curve in Figure 7.5 is characterized by a series of peaks and troughs which are potentially the result of rainfall events and management practices throughout the year. This is, however, only representative of the general trend in water table levels in the study area. The impact of rainfall on each wetland is likely to be significantly different as a result of the variation in land use and hydrological management between sites and the characteristics of each wetland's catchment.

Although incomplete, the mean weekly rainfall curve incorporated in Figure 7.6, which is derived from daily data from the Metu (August to October), Sor and Dizi gauges, explains some aspects of the behaviour of the water table elevation. Specifically, the abnormally high level of rainfall during March and the low levels for April would appear to account for the recovery and the steep drop in the water table level during these months respectively. Under normal rainfall conditions, the gradual decline in water table elevation would arguably continue until a significant increase in rainfall occurred, usually during April, when the water table would rise again.

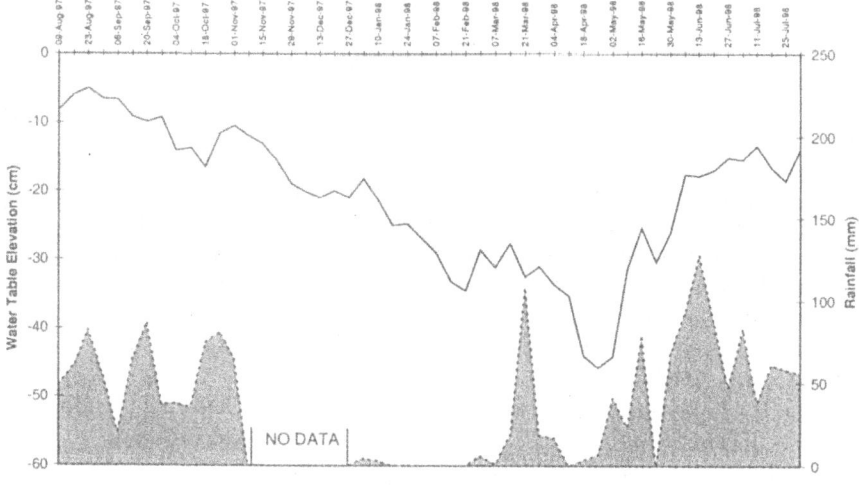

Figure 7.5 The general trend in water table levels during the study period

Variation between wetlands

Figures 7.6 and 7.7 show the mean weekly water table height of each wetland during the study period. This highlights the distinct difference between those sites which were currently undrained (Chebere, Tulube and upper Wangeneye) and those which were drained or degraded (Dizi, Anger, Bake Chora, Hurumu and lower Wangeneye), supporting the classification of wetlands on the basis of their development. From this it is clear that the currently undrained wetlands, apart from an obvious higher water level, also demonstrate more uniform water table elevations than the drained and degraded wetlands with lower and highly variable water table elevations. With respect to the currently undrained sites, Chebere (the near pristine wetland abandoned since 1992) generally maintains the highest water table throughout the year. This is followed by Tulube (abandoned some time in the late 1980s but regularly used for grazing) and then the undrained area at the head of Wangeneye wetland (abandoned since 1992).

The relationships between the drained sites of Bake Chora, Dizi, Anger, lower Wangeneye and the degraded site of Hurumu, appear to be much more complex. Basic descriptive statistics (Figure 7.8) confirm that on no occasion does the mean weekly water table height rise above the surface of the wetland. Furthermore, these sites exhibit a much greater dispersion in their absolute range of values over the year and this occurs over a lower range of water table heights than the currently undrained wetlands. This greater degree of inherent variation in the mean weekly water table elevation of the drained wetlands (in contrast to the currently undrained wetlands) can be explained as a result of three factors:

1. the site specific hydrological conditions;
2. changes in the number of operational dipwells which contribute to each weekly mean;
3. the actual human interaction and influence on the wetland's hydrological regime, which may include any activity ranging from drain excavation to land clearing (whose significance for the hydrological regime is discussed further in Chapters Eight and Nine).

In contrast to the three currently undrained wetlands (Chebere, Tulube and upper Wangeneye), the mean water table elevations in the drained and degraded sites also appear to be more variable from week to week, as shown by the greater variability of the curves (Figure 7.6). This suggests that rainfall, which can be considered spatially more even on a weekly time scale and which constitutes the common variable to all the study wetlands, is not the only factor which produces this irregular pattern in the water table data. Moreover, it suggests that 'pristine' wetland areas exhibit a slower hydrological response to rainfall events. For example, the recovery of water table levels at the end of October in response to rainfall is seen to be very steep in Dizi, Anger, Bake Chora and Hurumu.

In contrast, the impact of this rainfall in Tulube and upper Wangeneye is represented by a small, gradual increase to a peak which occurs approximately one week later than the drained and degraded sites. Furthermore, the impact on

The Hydrology of Valley Bottom Wetlands 127

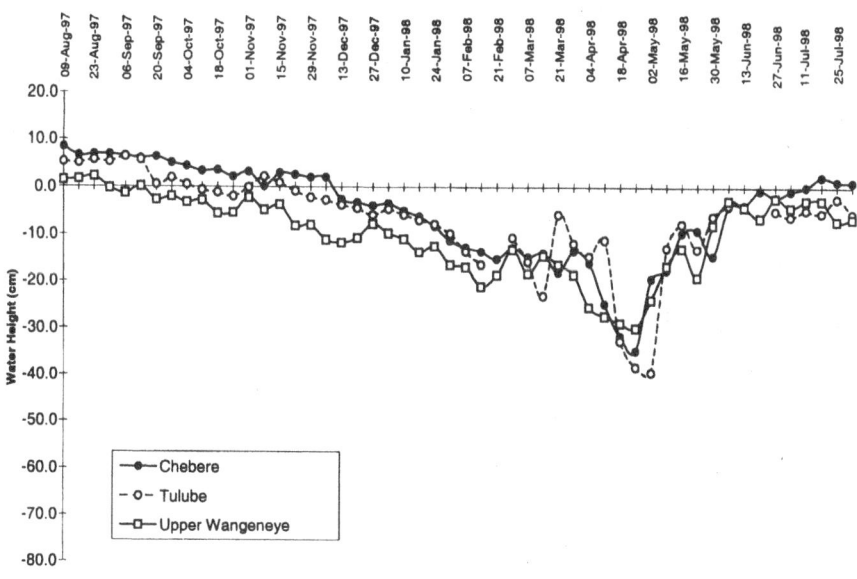

Figure 7.6 Mean weekly water table elevation in the currently undrained study wetlands (August 1997 – July 1998)

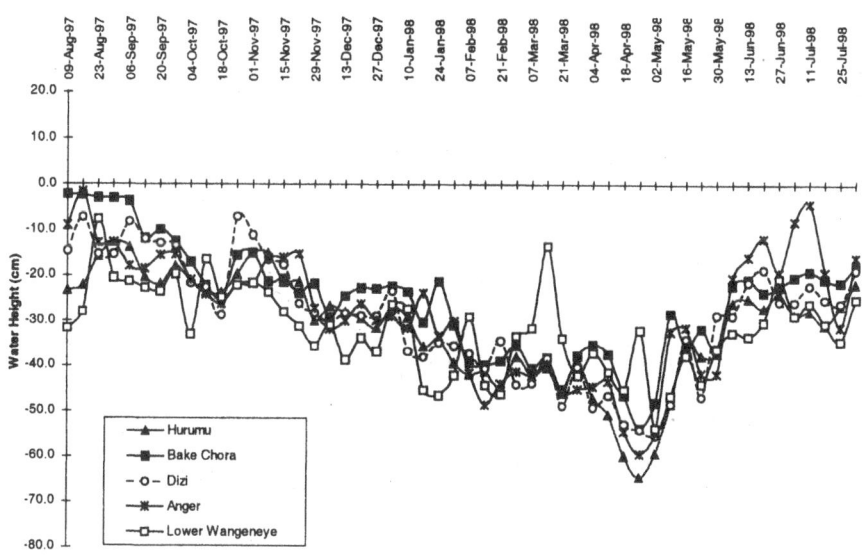

Figure 7.7 Mean weekly water table elevation in the currently drained and degraded study wetlands (August 1997 – July 1998)

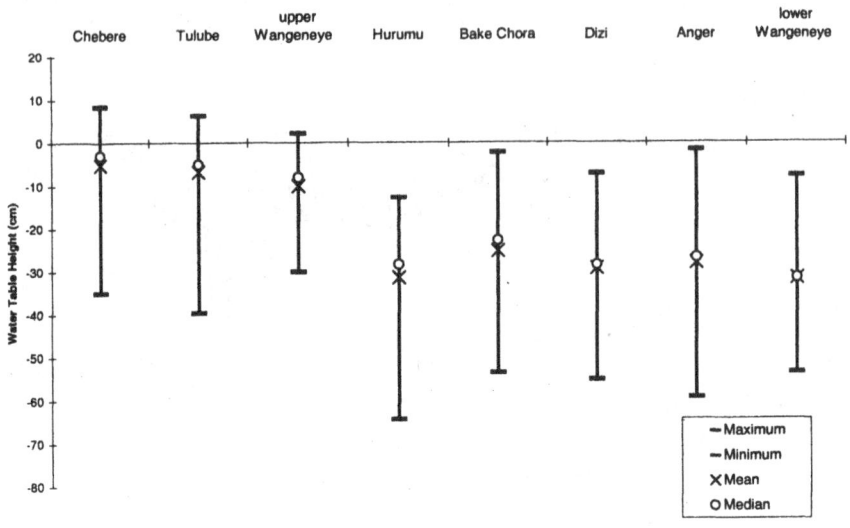

Figure 7.8 The range of mean weekly water table elevations recorded at each study wetland (August 1997 – July 1998)

Chebere wetland is barely noticeable. It is, therefore, more likely that development activities (such as drainage and cultivation) both in the wetland and the surrounding catchment area (which affect soil and vegetation) contribute to the observed erratic peaks in water table by interfering with the effects of rainfall.

Variations within wetlands: a wetland water table typology using cluster analysis

To explore the variation between different wetlands further, and to identify the impact of specific wetland land-use on water-table behaviour, the statistical technique of cluster analysis was employed. By compiling the year's data from each dipwell and analyzing the similarity between each, the cluster analysis identified four groups of dipwells, within which broadly similar water-table behaviour was represented.

Initial examination of the clusters of dipwells classified by this method revealed several patterns. In particular, the first cluster is composed almost exclusively of those dipwells representing areas of natural wetland vegetation mostly in undrained areas, whilst the second is dominated by dipwells representing wet areas within drained sites. Dipwells in the third cluster appear to represent well drained areas and finally those in cluster four are those which exhibit consistently low water levels. Table 7.1 summarizes the hydrological characteristics of each group based on a detailed analysis presented in Dixon (2000).

Table 7.1 Summary of the hydrological characteristics of each cluster of dipwells

Cluster	Water-table characteristics	Location characteristics	Number of constitutent dipwells in each wetland
Group 1 Max: 32 Min: -78 Mean Lvl: -7 Mean SD: 12.9	• Slow, gradual decline in elevation during dry season, almost full recovery during rains.	• Dipwells represent saturated areas, almost exclusive to undisturbed locations.	Chebere (11) Tulube (10) Wangeneye (3) Anger (4) Bake Chora (4) Dizi (1)
Group 2 Max: -15 Min: -74 Mean Lvl: -22 Mean SD: 14.2	• Weekly undulating rise and decline during dry season, extremely variable after February, steep recovery during rains.	• Mostly disturbed sites, also on the periphery of undisturbed sites.	Dizi (5) Wangeneye (4) Bake Chora (5) Hurumu (3) Chebere (1) Tulube (2)
Group 3 Max: 4 Min: -87 Mean Lvl: -30 Mean SD: 13.8	• Weekly undulating rise and decline (at lower elevations than Group 2) throughout whole season.	• Disturbed, heavily-drained areas.	Dizi (6) Anger (5) Bake Chora (4) Hurumu (3) Wangeneye (2)
Group 4 Max: -5 Min: -88 Mean Lvl: -43 Mean SD: 12.3	• Variable, low elevation throughout the year, limited recovery during rains. No uniform response to rainfall.	• Dipwells located in disturbed areas with deep drains (elevated wetland surface).	Anger (3) Hurumu (2) Wangeneye (2)

Table 7.2 Number of weeks with surface water at each dipwell

Dipwell No.	1	2	3	4	5	6	7	8	9	10	11	12	Mean
Chebere	1	15	13	30	18	25	30	26	22	13	9	18	18.3
Tulube	19	29	6	9	8	21	2	18	7	8	34	21	15.2
U. Wangeneye	1	2	0	7	15	24							8.2
L. Wangeneye							3	0	0	1	0		0.8
Dizi	0	1	1	1	0	0	0	1	1	0	2	0	0.5
Bake Chora	13	1	0	13	2	0	2	3	5	5	0	1	3.8
Anger	24	30	0	0	1	0	0	7	11	0	3	0	6.3
Hurumu	0	0	0	4	0	1	0	0	0				0.6

In addition to the cluster analysis, each dipwell was also classified according to its surface wetness, i.e. the length of time during which the water-table level is at or above the wetland surface (Table 7.2). By plotting this classification with the results of the cluster analysis on a series of wetland maps the spatial variability in water-table characteristics could be viewed in the context of land use and the effects of human interaction with each wetland (Figure 7.9).

These maps clearly highlight the wetter areas of each wetland, in particular drawing attention to Chebere, Tulube and upper Wangeneye which experience surface flooding for longer periods than other sites. The surface wetness index, however, does not appear to be evenly distributed throughout these wetlands and this is particularly clear in Tulube where the middle of the dipwell transect is drier than the top or the bottom. On the currently drained sites such as Bake Chora and Anger, surface wetness is highly spatially variable.

The maps also show that there is no clear relationship between the extent of surface flooding and the classification of hydrological behaviour throughout the year. For example, whilst all but one of the dipwells in Chebere constitute group 1 dipwells according to the cluster analysis, the wetness index is spatially variable. Similarly in Tulube, the range of wetness values is greater yet only two of the dipwells are not of group 1. The factors affecting surface saturation would, therefore, appear to be independent of those which determine the response of the wetland water table to any hydrological inputs.

Discussion of the water table typology

The key finding of this typology is that differences in water table behaviour do not appear to be specific to individual wetlands. Instead, the differences appear to reflect the existence of specific wetland sub-units which are ubiquitous throughout the study wetlands. The occurrence of these sub-units are clearly associated with a particular hydrological regime which appears to be linked to land use characteristics or the presence of rocks or springs.

A second key aspect of this water table typology is the difference in the behaviour of the water table between abandoned, pristine and cultivated areas within each wetland, which as discussed earlier, is independent of the degree to which water rises above the surface of the wetland. The typology suggests that the relationship between wetlands is complex and the differences are based on the abundance of particular wetland sub-units (as represented by the location of dipwells) which contribute to the mean weekly water table value.

Although the utilization strategy may have a direct influence on the behaviour of the water table throughout the season, the secondary effects of repeated utilization, in particular drainage and cattle grazing, include the physical alteration of the soil structure and subsequently the hydraulic conductivity of the soil. The scenario in undrained wetlands is one where soils exhibit higher hydraulic conductivities as a result of their higher organic matter content and lower bulk density (Belay Tegegne, 1998). Consequently, there would be less spatial variability in water table levels throughout the wetland producing a more homogeneous hydrological unit which is unaffected by artificial drainage.

The Hydrology of Valley Bottom Wetlands   131

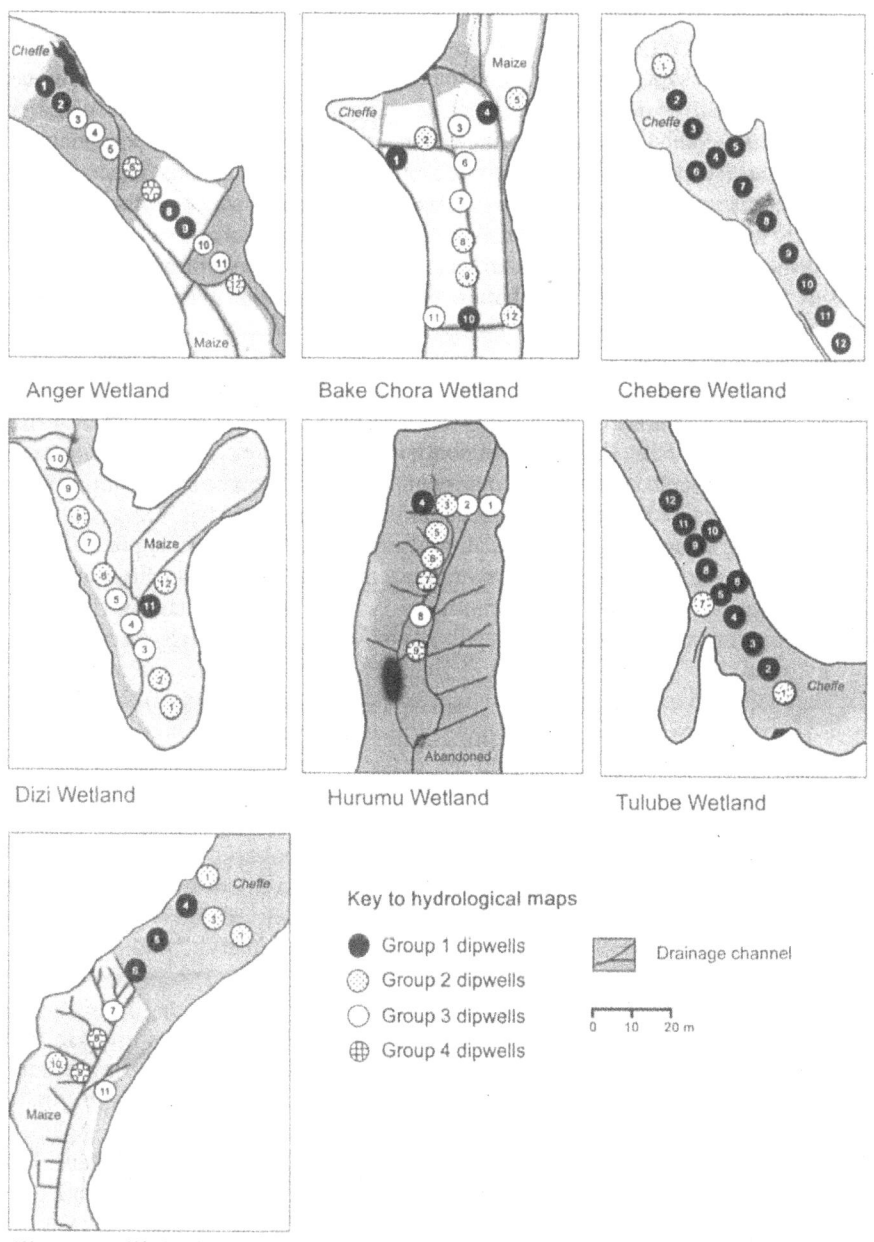

Figure 7.9 The spatial distribution of wetland hydrological characteristics in the study wetlands

Meanwhile, the presence of a natural outflow from these wetlands promotes the slow and regulated release of water from these areas.

Such a scenario does in fact appear to be the case for both Chebere and Tulube, where most of the dipwells fit into the group 1 category. In addition, the occurrence of group 1 dipwells in other wetlands points towards the existence of sub-units which display similar physical properties to these undrained sites. In drained or degraded wetlands these similar sub-units are typically located within, or directly adjacent to, undrained areas of natural vegetation.

Conversely, the soils of degraded wetlands are likely to be characterized by a low organic matter content and more compacted soil as a result of farming activities and the mineralization of the soil. These effects have been reported during studies on upland valley swamps in Uganda which have undergone drainage (Denny and Turyatunga, 1992). In addition, several researchers have suggested that *dambo* soils are degraded as a result of drainage, cultivation and overgrazing (Turner, 1986; Bullock, 1992b). Commonly reported soil impacts include erosion, a reduction in organic matter content, desiccation of soils and reduced infiltration capacities during rainfall events (Roberts and Lambert, 1990).

In view of these effects, degraded wetlands which have undergone drainage would be expected to have lower hydraulic conductivities and infiltration rates, which in turn will increase the accumulation of water at the wetland surface. This surface ponding effect was observed in several wetlands although particularly at Anger (dipwells 7, 8 and 9) and throughout Hurumu during the rainy season. In graphical terms, the effect on water table behaviour of these circumstances would be characterized by a curve which shows little or no seasonal response to rainfall, arguably because recharge is affected by reduced infiltration rates and the effects of surface runoff and evaporation.

In those areas where drainage and cultivation is ongoing the picture is less clear. Group 2 and 3 are abundant throughout the wetlands and both groups exhibit fluctuations from week to week. Although group 2 dipwells occur in areas where cultivation and drainage are taking place (Figure 7.9), they are also located in areas where there is relatively little human interference (particularly Tulube and upper Wangeneye) and on the periphery of group 1 areas, suggesting some similarity in terms of the hydrological regime of these areas. Group 3 dipwells, however, are exclusively located in those areas undergoing drainage and the majority are also characterized by a water table which never reaches the surface of the wetland, whilst group 2 dipwells predominantly exhibit a surface water table for up to 10 weeks of the year (Table 7.2).

A comparison of the basic descriptive statistics for group 2 and 3 (Table 7.1) reveal that group 2 dipwells also have a higher mean water level (-22 cm) than group 3 dipwells (-30 cm). By relating the result of the dipwell typology to the actual land use characteristics of each wetland, there is clear evidence to suggest that group 2 dipwells are associated with wetter conditions more akin to the characteristics represented by group 1 dipwells, and that the presence of group 3 (and group 4) dipwells in each wetland is indicative of drier and potentially better drained areas. This relationship is summarized in Figure 7.10.

The Hydrology of Valley Bottom Wetlands 133

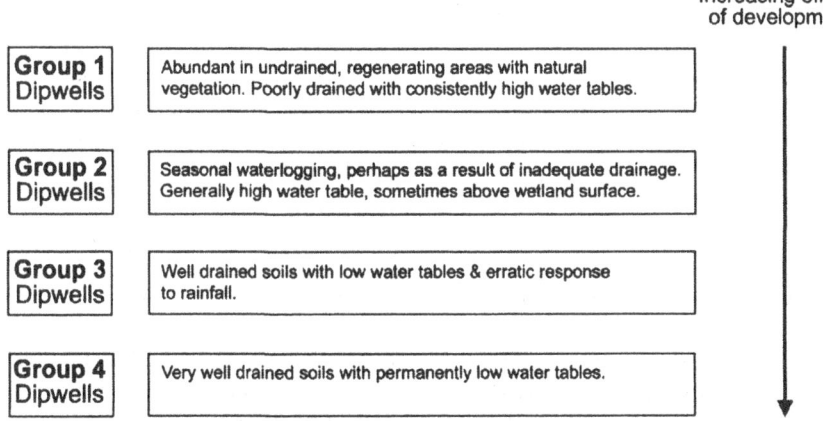

Figure 7.10   The relationship between dipwell groups

Although the past and present land use goes some way to explaining the differences in hydrological behaviour of wetland areas, another cause of variation is topography. The gradient of the slope towards the outflow and the uneven wetland surface, characterized by hollows and raised areas, can contribute to the mean water level and the fluctuations recorded by dipwells. Within the space of only a few metres the surface of the wetlands may vary as a result of the influence of dense vegetation as in the case of Chebere, or because of the uneven redistribution of soil which follows ploughing (Bake Chora and Wangeneye). This will influence both the speed and direction of surface runoff. Furthermore, the uneven surface of each wetland can facilitate the accumulation of water in certain areas and if the soil is relatively impermeable (as is possible in heavily utilized or degraded sites) the effect may be similar to that of a perched water table, the repercussions of which, are a misrepresentation of water table elevation as recorded by the dipwells.

This may be the case for dipwells 8 and 9 in Anger wetland where the water table was recorded above the surface for a significant period throughout the year, whereas water levels in the adjacent dipwells did not rise above the surface at any time. During field visits to Anger wetland in June and July 1997 the areas represented by dipwells 8 and 9 were observed to be flooded and the maize crop severely damaged as a result, despite apparently well drained areas on either side. This could be explained by the relatively flat surface of this part of the wetland, which exhibits a gradient towards the outflow of only 1.1°.

Similarly during the wet season of 1998 surface waterlogging of Hurumu wetland, particularly around dipwells 1 to 6, was a commonly observed feature even though hydrological records indicate that these dipwells are among the driest of the study wetlands. It is unclear, however, the extent to which this behaviour is a product of topographical variation, original soil characteristics or, as mentioned above, the influence of various land use factors (e.g. grazing, ploughing).

As an indicator of water table behaviour, both the land use and topography of the specific area should, therefore, be considered influential. Land use in particular, has a significant effect on the recharge and subsequent behaviour of the water table whether this is a result of continuous drainage and cultivation, or the trampling effects of cattle. The major determinant of water table variation between sites, however, appears to be the past and present influence of drainage and the impacts which this and the physical process of cultivation has on wetland soil (and subsequently hydrological) characteristics.

Saturated hydraulic conductivity in the wetlands

Measurement of K_{sat} was carried out in Anger, Bake Chora, Dizi and Hurumu, but not in Chebere and Tulube as K_{sat} values were high to such an extent that the recovery rate of water elevation was immediate. For comparative purposes, the K_{sat} results of a more recent study (Conway, 1999) are also included in Table 7.3.

The FAO (1963) classification shown in Table 7.3 places the hydraulic conductivity of the wetland soils between the categories of 'very slow' and 'very rapid', the latter describing those sites where instant recovery was observed. The range of K_{sat} values recorded are also indicative of soils ranging in texture from heavy clay (<0.05 m day^{-1}) to sandy loam (1.5 - 3.0 m day^{-1}) (FAO, 1979). Of the sites where K_{sat} measurements were initially recorded, Anger was found to have the most variation in its range of values, followed by Dizi, Bake Chora and then Hurumu.

This wide variation in K_{sat} reflects the presence of wetland sub-units which relate to land use and to some extent, topography, as discussed previously. Anger wetland exhibits the most diverse range of K_{sat} values reflecting the variability in the current and historical utilization of the instrumented part of the wetland. In view of the dipwell typology, those areas with high K_{sat} values (dipwells 1, 2, 8 and 9) correspond to group 1 sub-units, which exhibit relatively high water table elevations throughout the year. The data from Conway (1999), however, conflicts with that of the original Anger study and to a lesser extent with data collected during 1997 and 1998 at the other wetlands.

Several explanations can be put forward for these discrepancies. First, the pump tests were carried out at different times of the year: the first set during the height of the wet season and the second set towards the end of the dry season. Although theoretically K_{sat} should be a constant value, the influence of drought conditions and animal grazing could potentially lead to a reduction in K_{sat} through soil desiccation and / or compaction (as has been reported with regards to *dambo* soils, Roberts, 1988; McFarlane and Whitlow, 1990).

Alternatively, anthropogenic disturbance of the wetland soils through activities such as ploughing may increase the pore space in the wetland soil and consequently increase K_{sat}. Furthermore, depending upon the water table level at the time of the pump tests, recovery of the water column will be influenced by the K_{sat} of the particular soil horizon where the inflowing water comes from (K_{sat} may

Table 7.3 Classification of K$_{sat}$ values in each dipwell according to FAO (1963)

Dip.	Chebere Jul 97	Chebere Apr 99	Tulube Jul 97	Tulube Apr 99	Hurumu Jul 97	Hurumu Apr 99	Bake Chora Feb/Apr 98	Bake Chora Apr 99	Wangeneye (upper: 1-6, lower: 7-12) Apr 99	Anger Jul 97	Anger Apr 99	Dizi Jul 97	Dizi Apr 99
1	vr	-	vr	-	s	m	vs	vr	-	vr	vr	vr	vs	s
2	vr	-	vr	-	vs	s	vs	-	-	vr	vr	vs	vs	-
3	vr	-	vr	-	vs	r	s	m	-	-	vs	vs	vs	m
4	vr	vr	vr	-	vs	-	m	m	-	-	s	vs	vr	-
5	vr	vr	vr	-	vs	-	s	m	-	vr	s	-	vs	-
6	vr	vr	vr	-	vs	s	vs	-	-	vr	vs	-	vs	s
7	vr	vr	vr	-	vs	-	vs	-	-	-	r	-	vs	-
8	vr	vr	vr	-	vs	vs	vs	-	-	-	vr	vs	vr	vr
9	vr	vr	vr	-	vs	r	vs	s	-	-	vr	vs	vs	m
10	vr	vr	vr	-			vs	mr	-	-	vs	m	s	-
11	vr	vr	vr	-			s	-	-	-	vs	vs	m	-
12	vr	vr	vr	-			-	-	-	-	-	-	vs	-

(Incorporating data from Conway (1999))

vs:	Very slow	<0.2 m day^{-1}	clay, silty clay
s:	Slow	0.2 - 0.5 m day^{-1}	clay loam, sandy clay loam
m:	Moderate	0.5 - 1.4 m day^{-1}	light clay loam, silt loam
mr:	Moderately rapid	1.4 - 1.9 m day^{-1}	very fine sandy loam, loam
r:	Rapid	1.9 - 3.0 m day^{-1}	sandy loam
vr:	Very rapid	>3.0 m day^{-1}	loamy sand, sand, coarse sand
- :	missing values (missing dipwells, silted dipwells)		

vary considerably between horizons). During the dry season these will be the deeper horizons which tend to have higher clay fractions and higher bulk densities than the more organic rich upper horizons.

In contrast to Anger, the dipwell transect in Hurumu wetland (degraded status) represents a relatively homogeneous, abandoned land unit, which would explain the narrow range of K$_{sat}$ values recorded throughout the wetland which do not exceed 0.24 m day^{-1}. The water table typology, however, suggests a fairly wide range of water table behaviour and dipwell 4, whose K$_{sat}$ value was recorded as 0.075 m day^{-1}, represents a group 1 sub-unit usually associated with soils with very high K$_{sat}$. Nonetheless, other areas within Hurumu, which exhibited extremely slow K$_{sat}$s of less than 0.005 m day^{-1} (dipwells 2, 8 and 9) correspond to the drier areas identified in the typology.

As in Hurumu, the observed variation in K$_{sat}$ values in Dizi does not fully correspond to the distribution of water table sub-units within the wetland, although overall, K$_{sat}$s were found to be very slow as expected in view of the land use.

Whilst the group 1 dipwell (11) in the wetland does exhibit a higher K_{sat} than the majority (1.2 m day^{-1}), an adjacent group 3 dipwell (4) exhibited immediate recovery of the water table during the pump test, although this may have been influenced by its proximity to the nearby stream channel (Figure 7.9).

In terms of drainage, Bake Chora is similar to Dizi in that it is fully cultivated and the wetland is dissected by numerous drainage channels. Throughout most of the wetland K_{sat} measurements tend to fall into the category of 'very slow' for the February/April 1998 studies (the mean value of two separate investigations was used) confirming the predominance of drier sub-units. The study by Conway (1999) although incomplete, estimates K_{sat} values as ranging from 'moderate' to 'very rapid' in those areas identified by the dipwell typology as wetter, in particular dipwells 1, 4, 5 and 10.

Although the K_{sat} measurements are broadly consistent with those expected in view of the water table results, there are clear instances where the results of each K_{sat} study conflict and do not correspond with the characteristics expected of a specific sub-unit. Apart from the mechanisms which disturb the soil profile as mentioned above, the proximity of drainage channels to dipwells can influence the degree of saturation in the adjacent wetland soil. The concentration of interflow from the contributing area above the drain (the accumulation of subsurface flow above the groundwater table) can cause an increase in the water table elevation immediately next to the channel. As a result, the measurement of K_{sat} in these areas may be representative of a different depth of the soil profile.

There are also several limitations of the pump test method employed. The data represents the measurement of K_{sat} on only one occasion during the study period, when in fact the variability of individual readings taken in the field can be large, even between points close to each other. Furthermore, this variation can be more significant in the top 45 cm of the soil, where management can have a greater effect on soil properties (Landon, 1991).

The use of dipwells rather than an augured hole may also have been a source of error since the speed of elevation recovery will be slower in the dipwells where the perforated PVC lining hinders recharge. Additionally, as the water table decreases naturally, the closed bottom of the dipwell traps residual water which may be misinterpreted as the actual elevation. If this is removed from the dipwell column during a pump test, recharge will not occur and this may have been interpreted as an extremely slow K_{sat}. This is, however, unlikely given that any drop in the water table elevation beyond the base of the dipwell tube was characterized by the retention of sediment and mud rather than a water column, which could not be pumped from the dipwell. Under these circumstances, dipwell was ignored and replaced.

In summary, there is a need to replicate pump tests at each dipwell and at different times throughout the year in order to improve the validity of the data and give some indication of the variability within the soil profile throughout each wetland. Although these investigations tend to suggest that slower K_{sats} are a characteristic of those wetlands which are currently drained or degraded, the data is far from conclusive.

Water chemistry

The monitoring of several water quality parameters was undertaken to establish the seasonal changes in wetland nutrient levels and the impact of wetland development on the nutrient status of wetlands in the study area. In addition, this aspect of the research also sought to establish the dynamics of water quality in each wetland through the monitoring of water chemistry in both the inflow and outflow of each wetland.

Measurements of pH, electrical conductivity, phosphate and nitrate were made at the top and bottom of each wetland, during the study period and up to July 1999 (owing to equipment failure during the initial study period).

pH levels in the wetlands

The mean weekly pH level shown in Figure 7.11 is derived from the mean of all seven wetlands. This in turn, is derived from weekly recordings taken from the top and bottom of the each wetland. Although it suffers from long gaps in the record particularly during the 1997 – 1998 season, a pattern of increasing alkalinity during the drier months of the year (November to April) followed by a decline during the wet season (May to October) is evident. This represents a move from neutral to acidic conditions following the onset of heavy rains.

In terms of the variation in pH levels between wetlands, the data suggests that there is little difference (Figure 7.12) although Chebere exhibits a slightly narrower range of variation during the period and Wangeneye the largest (possibly a result of the more rapid runoff associated with Wangeneye's adjacent road).

With regards to variation within each wetland, Figure 7.13 shows the mean of the weekly pH values recorded from the top and the bottom of each wetland. Again, there is no clear relationship over time between pH recorded at the top and the bottom of the wetland. During the data period the mean level at the top exceeded that of the bottom on 32 occasions and vice versa on 35 occasions, although on the evidence presented in Figure 7.26 this situation appears to occur at random. In most cases the differences recorded between the top and bottom of each wetland are negligible and rarely amount to greater than 0.5.

In view of the short distances between the top and bottom of the sampling area which are directly linked by natural or artificial channels in the developed and degraded sites, the lack of any variation is not particularly surprising. In Tulube and Chebere, which do not possess such channels, the opportunity for the leaching of minerals between the top and bottom of the wetlands is, however, potentially greater. The only obvious behavioural trend in pH levels is the influence of rainfall throughout the study wetlands, where acidity in the wetland runoff rises during the rainy season (June – November 1998) and then decreases after the rains subside (December 1998 – May 1999). This may be indicative of a period of leaching and flushing of organic matter, pesticides or fertilizers from the catchment and the wetland itself.

138   *Indigenous Management of Wetlands*

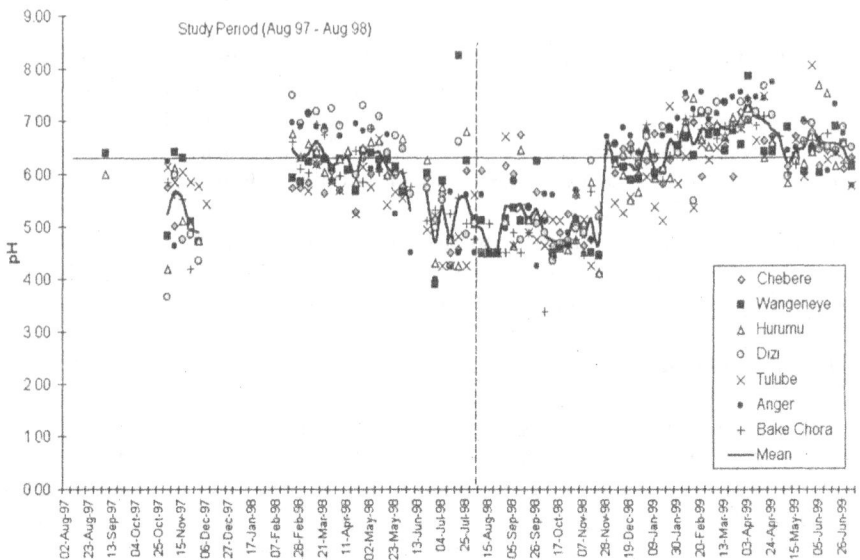

Figure 7.11   Mean weekly pH levels in the study wetlands

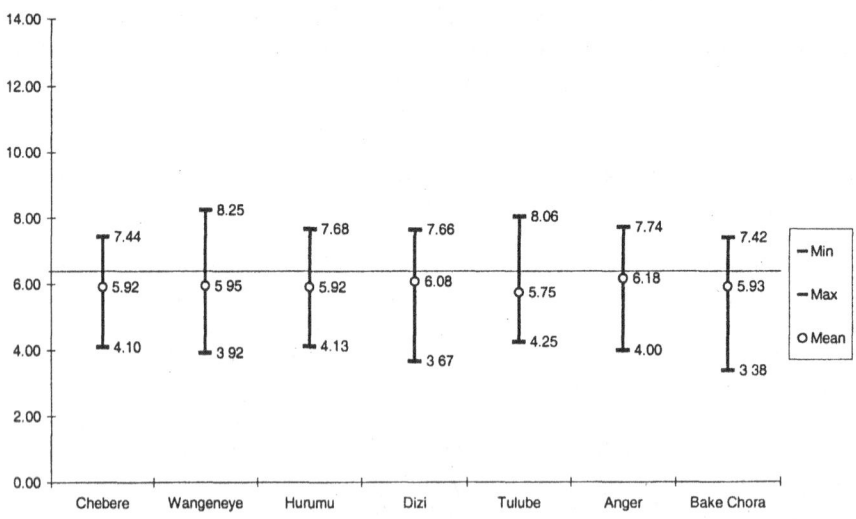

Figure 7.12   Range of pH values in each study wetland (August 1997 – July 1999)

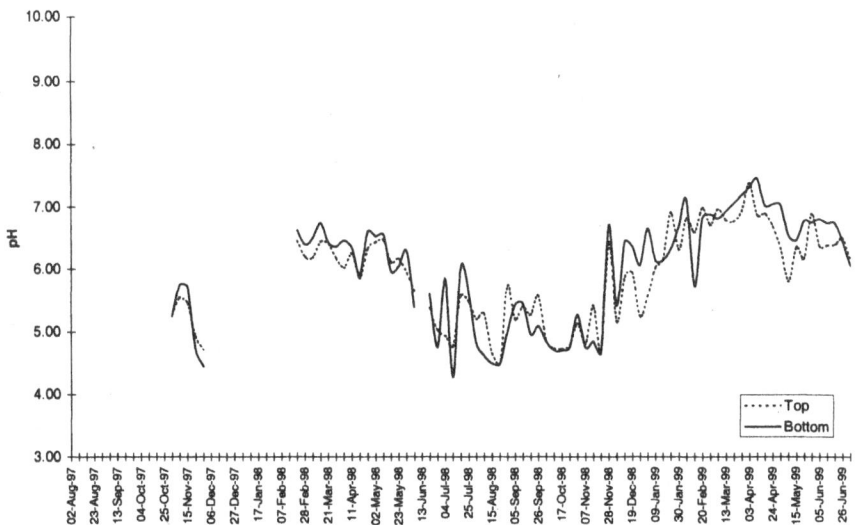

Figure 7.13   Mean weekly pH levels recorded at the top and bottom of each wetland (August 1997 – July 1999, including periods of missing data)

Electrical conductivity

The mean weekly electrical conductivity (EC) level for all the wetlands is shown in Figure 7.14. This shows that over the extended period electrical conductivity remained relatively low. These observed values are indicative of clean water with a low content of dissolved salts, which does not appear to exhibit any seasonality in terms of response to rainfall.

Figure 7.15 also suggests that there is no obvious difference between EC at the top and bottom of each wetland. Whilst there is a noticeable increase in levels after February 1998, this is more likely to be a result of the change in monitoring equipment used during that period rather than any real change in EC, especially as hydro-meteorological conditions varied little during this dry period. If flushing was to occur, it would most likely happen with the onset of the main rains, which in 1998 occurred much later than usual (May rather than April). Although it is possible that human activity in the wetlands (such as drainage) could cause the release of dissolved salts from the wetland, drainage itself is usually carried out much earlier in the season.

In each wetland there does not appear to be any seasonal trend and no discernable difference between drained or undrained, headwater or mid-valley wetlands, raising questions about the influence of drainage on EC levels. In summary, the data suggests that dissolved salt concentrations in the wetland outlets remain relatively similar to the concentrations recorded in the inflows throughout the year despite human disturbance.

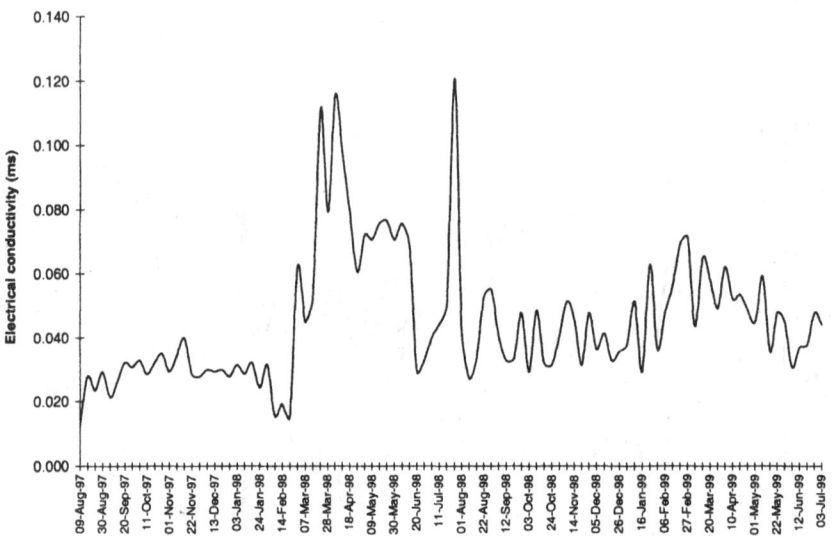

Figure 7.14 The mean weekly electrical conductivity in the study wetlands (August 1997 – July 1999)

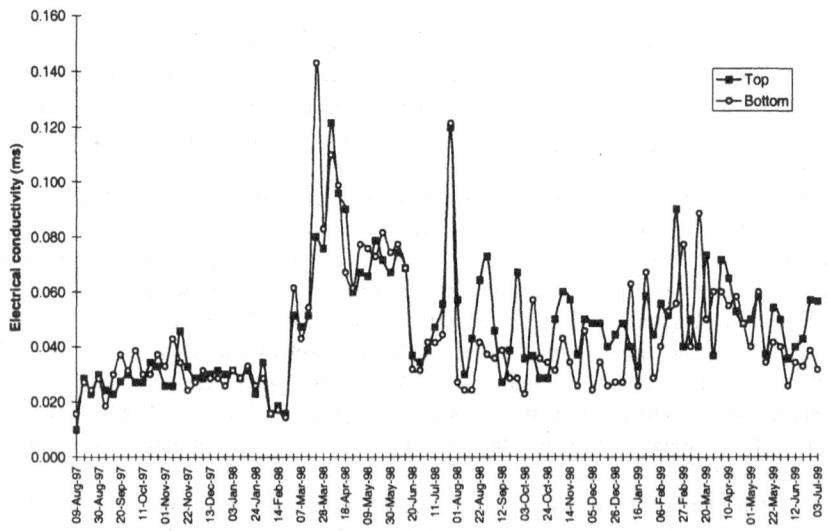

Figure 7.15 The mean weekly electrical conductivity at the top and bottom of the study wetlands (August 1997 – July 1999)

Nitrate and phosphate concentrations in the wetlands

Figure 7.16 shows the monthly mean of all the nitrate and phosphate recordings taken at each site, for each wetland. This gives a broad indication of the trend in nitrate and phosphate behaviour in the study wetlands during and after the study period. It suggests that first, the concentration of phosphate within the wetland drainage system tends to be higher than the nitrate concentration and that this reaches a peak between October and November 1998. Secondly, it suggests that nitrate and phosphate concentrations do not appear to follow the same seasonal trends.

Correlation analysis of the mean values recorded at the top and bottom of each wetland, shows that phosphate concentrations at the top and bottom show no clear relationship with each other (0.38, which is not statistically significant at a 0.05 level). In contrast, nitrate concentrations from the top and bottom do show similar seasonal trends, having a higher correlation coefficient of 0.60 (significant at the 0.05 level).

In terms of the difference between wetlands, the correlation matrices presented in Table 7.4, indicate strong correlations between the trend in nitrate and phosphate concentrations between several wetlands. In particular, the positive correlation in phosphate concentrations between Tulube and Wangeneye (0.52) and Tulube and Bake Chora (0.57), are statistically significant. In terms of nitrate concentrations, Tulube and Bake Chora among several other sites, also show a strong, positive correlation with each other (0.83). The data does not, however, appear to reflect any similarities or differences between sites in terms of their development status or hydrological classification.

The data sets for both nitrate and phosphate show a considerable range of variation in terms of differences between each wetland and between the concentrations recorded at the top and the bottom. Despite the lack of any clear seasonal trend, however, several key conclusions can be drawn from the data.

First, the results shown in Table 7.5 suggest that nitrate levels are less variable than phosphate, which in contrast, show a much higher degree of spatial and temporal variability. The fact that nitrate concentrations demonstrate stronger positive correlations between sites suggests that they are less likely to be influenced by site specific practices such as natural or artificial fertilizer application or pollution (unless these are carried out at the same time on all the sites). In contrast, the almost random variation which is characteristic of the pattern of phosphate concentrations in each site, suggests point source contamination of phosphate.

In addition, the relationship between chemical concentrations at the top and bottom of each wetland is ambiguous. This is the case between the wetland sites and for both the nitrate and phosphate measurements taken at each. For example, Tulube and Bake Chora exhibit higher nitrate concentrations at the wetland inflows throughout the study period whereas at the other sites the results are variable. For phosphate concentrations, however, Chebere, Bake Chora and Hurumu have higher concentrations at the wetland inflows.

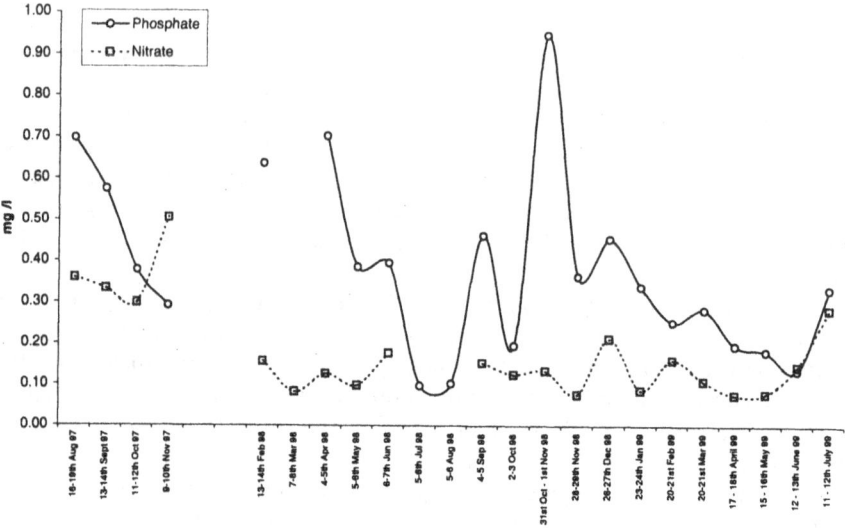

Figure 7.16 Mean monthly nitrate and phosphate levels recorded in the study wetlands (August 1997 – July 1999)

Table 7.4 Correlation matrices for mean phosphate and nitrate concentrations

a) Phosphate Concentrations

	Wangeneye	Chebere	Dizi	Anger	Bake Chora	Tulube
Chebere	0.32					
Dizi	-0.01	-0.16				
Anger	-0.07	0.39	0.15			
Bake Chora	0.27	0.34	-0.20	-0.02		
Tulube	0.52	0.09	-0.05	-0.06	0.57	
Hurumu	-0.08	0.05	-0.01	0.38	-0.10	-0.12

b) Nitrate Concentrations

	Wangeneye	Chebere	Dizi	Anger	Bake Chora	Tulube
Chebere	0.10					
Dizi	-0.04	-0.09				
Anger	0.00	0.60	0.08			
Bake Chora	0.15	0.52	-0.53	0.41		
Tulube	0.35	0.57	-0.51	0.40	0.83	
Hurumu	-0.05	0.20	0.23	0.08	-0.02	-0.24

Table 7.5 Summary details of chemical concentrations recorded at each wetland

	Nitrate				Phosphate			
	n	Mean	St Dev	CV	n	Mean	St Dev	CV
Chebere	21	0.05	0.05	100	22	0.70	0.5	75
Tulube	19	0.13	0.16	122	21	0.36	0.6	165
Wangeneye	21	0.14	0.2	142	22	0.24	0.3	121
Bake Chora	21	0.51	0.54	106	20	0.57	0.5	92
Hurumu	21	0.17	0.15	89	20	0.27	0.4	134
Dizi	21	0.07	0.04	53	20	0.26	0.2	90
Anger	20	0.18	0.08	47	20	0.21	0.4	168

n number of records
CV coefficient of variation

Finally, of all the sites, Bake Chora stands out as the most chemically variable. It exhibits consistently higher chemical concentrations at the top of the wetland rather than the bottom, significantly higher nitrate levels than all other sites and, relatively extreme fluctuations in phosphate concentrations from one month to the next.

Possible causes of the observed nitrate and phosphate patterns

Nitrate and phosphate concentrations within each wetland will be influenced by the level of application of artificial and natural fertilizers and any drainage activities which would lead to increased leaching into water channels (Ward and Robinson, 1990). The impact of these activities is likely to be increased concentrations in runoff, depending on the degree of human disturbance. Although artificial fertilizers are not being applied in any of the study wetlands or their surrounding catchment areas, natural fertilizers such as plant residue and animal manure are being utilized on the interfluves of all the sites with the exception of Hurumu (pers. comm. Afework Hailu, 1998). Furthermore, cattle contribute directly to the nutrient status of each wetland during ploughing activities or when they are allowed to graze, outside the growing season. The variation in the results, however, suggest that this contribution is inconsequential: cultivated wetlands show as much variation during the monitoring period as those uncultivated.

The impact of wetland drainage on nitrate and phosphate levels is also unclear. Bake Chora wetland shows several increases in phosphate concentrations which correspond to periods of drain excavation and ploughing (September and November 1998). These increases, however, were recorded at the top of the wetland, above the area of disturbance and, therefore, cannot have been directly affected by this activity. Chebere also demonstrates high concentrations at the top of the wetland between April and June 1998 which correspond to periods in which ploughing occurred on the valley sides.

Whilst nutrient levels remain highly variable in drained and undrained headwater wetlands, however, mid-valley drained wetlands exhibit a much narrower range of variation throughout the monitoring period. This is indicative of a situation where to a large extent, catchment runoff bypasses the wetland as it is rapidly conducted into a central channel where flow is maintained throughout the year. Consequently, these wetlands show less chemical variability than those which receive their water from springs and runoff within a small catchment, where rainfall itself is a much more localized event. The overall implication is that where longer water residence times are maintained, nutrient levels within the wetland are likely to be higher.

Another potential explanation for the observed peaks in nutrient load lies in the use of wells and springs for washing and bathing at the head of the wetlands. The washing of clothes with soap would explain the relatively high concentrations of phosphate recorded at the headwater spring of Bake Chora wetland, where this activity was observed on several occasions. Furthermore, inadequate drainage from the spring as a result of unmaintained drains would increase the residence time of the springwater, prolonging these high concentrations. The only other site where washing activities were observed in the headwater spring was Tulube, where results show that concentrations of both nitrate and phosphate were significantly lower than at Bake Chora. This difference can possibly be explained by the fact that for much of the study period, the spring at Tulube was drained by a better maintained channel, thereby reducing the spring's storage capacity and the residence time of water in it. An increase in nitrate concentrations can be attributed to the contamination of these areas by organic waste. Although fences are usually constructed around springs to protect them from cattle, these are not always well maintained.

Whilst these results suggest some degree of human influence on the behaviour of nutrient levels, given the observed variation, breaks in the data and the potential range of influences including natural fertilizers, cattle grazing and other human activities, it is hazardous to draw any firm conclusions. To put the data in perspective, the actual nitrate and phosphate levels recorded represent fairly low concentrations. Nitrate concentrations generally range between <5.0 mg/l in groundwater whose aquifer is located in regions where there is no farming or significant human influence, and up to 100 mg/l in drainage effluent (Price, 1996). Similarly, soluble phosphate concentrations vary from <10 mg/l in oligotrophic water to >100 mg/l in hypertrophic, nutrient rich water. In the study wetlands, phosphate concentrations ranged between 0.01 - 3.15 mg/l and nitrate ranged between <0.00 - 2.1 mg/l, results indicative of a nutrient poor environment.

From pristine to abandoned – the impact of drainage

Analysis of the data suggests that the wide range of hydrological differences observed between the 'pristine' site of Chebere and those which have been cultivated and subsequently abandoned, is evidence of the impact of agricultural utilization on wetland hydrology. In hydrological terms, utilization, in particular drainage and cultivation, has caused a shift from a wetland with a consistently high water table with very rapid hydraulic conductivity, to one with very different hydrological characteristics, i.e. a consistently low water table and very slow hydraulic conductivities.

The artificial lowering of the water table itself, however, cannot account completely for the degree of change which takes place. The hydraulic conductivity of Chebere is such that it would arguably require a very low density of drainage channels to conduct the groundwater away from the wetland. The cultivated sites, however, such as Wangeneye and Bake Chora, are characterized by an extensive network of drainage channels which are more suited towards the drainage of heavy soils with low hydraulic conductivities. Moreover, the data from these sites confirms the high spatial variability of water table elevation and hydraulic conductivity.

The main implication of these differences is that the physical characteristics of the wetland soil and the soil-moisture relationships appear to change significantly as a result of drainage and cultivation. The drainage network characterized by relatively closely spaced, deep drains, which exists on cultivated sites (e.g. in Wangeneye drains are often less than 5 m apart and up to 1 m in depth) probably represents an adaptation of the initial 'pristine wetland' drainage layout, which becomes inefficient once the drainage and cultivation process induce soil changes. As Trafford (1983) points out, once the hydraulic conductivity of a wetland soil falls below 0.1 m day^{-1} the required drain spacing for efficient drainage becomes so close that it cannot be justified on economic grounds. In addition, Landon (1991) argues that drainage requires critical planning if it is to be successful, where K_{sat} ranges between 0.1 and 1.0 m day^{-1}.

With repeated drainage and cultivation, the soil may continue to change in ways which reflect the land use of specific areas within wetlands and this may lead to an increase in the observed spatial variability in K_{sat} and water table behaviour. Efficient drainage, therefore, requires an intimate knowledge of the soil moisture conditions of different areas within each wetland and modifications to the drainage system made accordingly. Without such adjustments, the wetland may eventually become unsuitable for crop cultivation either because the drainage is too intensive and the recharge of the water table from runoff or rainfall is too slow (as in Hurumu), or the drainage is insufficient creating a waterlogged environment for crops (as in some parts of Anger, Dizi and Bake Chora). With overdrainage, the water table remains consistently low despite seasonal hydrological changes, exhibiting similar behaviour to the group 3 sub-units described earlier.

The potential changes in soil characteristics which can occur as a result of the drainage and utilization processes primarily include compaction and mineralization. The grazing of animals on wetlands during the year can result in the compaction of the surface soil horizons, increasing the bulk density and consequently reducing the pore spaces within the soil structure through which water can flow. Compaction is also exacerbated by the mineralization of soil brought about by repeated oxidation, which results in a decrease in organic matter content (Barrow, 1991). Furthermore, although the practice of burning crop

Table 7.6 Summary of the impact of agricultural utilization on the wetland hydrological regime

Development status	Water Table (GWT)	K_{sat}	Water chemistry
Pristine	Homogeneous GWT behaviour, high water table throughout year, gradual response to rainfall.	Very rapid throughout	Low variation in pH levels.
Fully cultivated	Spatially variable (wet and dry areas), flashy response to rainfall.	Variable	
Part drained	Uncultivated area exhibits pristine characteristics, drained area as above.	Very rapid in undrained area, variable in drained area	High variation in pH levels.
Degraded	Spatially variable although constantly low GWT dominates, flashy response to rainfall.	Very slow throughout	
Regenerating	Some variability in water table behaviour.	Very rapid throughout	

residues in the wetland may have the effect of concentrating nutrients in certain areas, it is also a destructive process which reduces the organic matter content at the soil surface and damages soil structure. As a consequence, the topsoil may be more sensitive to sheetwash, gully or wind erosion.

The expected impact of these processes, following the onset of wetland drainage and cultivation, would be a rapid decline in soil fertility as the organic matter in the topsoil is lost. Subsequently this would be followed by a more gradual decline in fertility as a result of the loss of soil from the wetland and the exposure of relatively impermeable clay soils. This degradation is, however, offset by the input of sediment and nutrients from the catchment (although this is dependant largely upon the land use characteristics of the catchment) and in addition, the flooding regime of the wetland itself which facilitates the depositional process. Ploughing is also an integral part of the wetland cultivation process and regarded by farmers as a means of improving the soil structure, albeit temporarily.

The result of these processes are evident in the characteristics of Hurumu and certain areas within Anger wetland which have been abandoned and are now currently utilized as grazing resources. Gully formation within the disused drain network of Hurumu is abundant and indicative of the steep slopes found in the surrounding catchment and the inability of water to infiltrate the heavily grazed, compacted wetland soil. Within these areas, however, there is some evidence of a process of regeneration taking place, during which the water table elevation increases and natural vegetation begins to recolonize the wetland surface, as has occurred in Tulube and upper Wangeneye wetlands. From a hydrological viewpoint, the causes of this regeneration are not clear although the rise in water table may be attributed to the collapse of drains and the low hydraulic conductivities of the degraded soils, facilitating waterlogged conditions. These in turn, provide the ecological conditions for the spread of pioneer wetland vegetation which climaxes in the growth of *cheffe*.

Cheffe itself, is characterized by an extensive network of rhizomes which increase the permeability and the organic matter content of the soil. Furthermore, the recolonization of *cheffe* impedes the flow of surface water through the wetland, thereby increasing residence time and sedimentation, which ultimately aids the restoration of fertile conditions. The proliferation of these conditions may take a number of years, as in the complete regeneration of 'pristine' characteristics akin to those at Chebere wetland, although recent indications from Tulube are that grazing can have a severe impact on the regeneration process.

Conclusions

The results of the hydrological monitoring programme has demonstrated that the installation of drainage networks in these valley bottom wetlands has led to a lowering of the water table level. Monitoring of the water table throughout the year has shown that those sites undergoing drainage show considerable spatial and temporal variation in their water table elevations in comparison to those wetlands classified as 'pristine' or 'regenerating'. While the active management of drains during the year may account for some of this variation, similar variability has been observed in the data for the 'degraded' site of Hurumu and in some areas of the 'regenerating' site of Tulube. Consequently, it has been suggested that the variable behaviour of the water table is dramatically affected by the secondary impacts of drainage (i.e. structural soil changes) combined with the effects of wetland cultivation.

The lowering of the water table results in the oxidation of organic matter, which is also depleted through the cultivation process. This causes a shrinkage of the soil, especially in the top layers, which also become susceptible to erosion. At the same time, the use of farm animals as part of the cultivation process and through grazing causes the compaction of the soil and the result is potentially a decrease in the hydraulic conductivity. As a result, drainage becomes less efficient and waterlogging can occur both in the soil profile and at the surface, where infiltration rates are reduced and less permeable soils from the lower horizons become exposed.

The spatial variability of the hydrology throughout each wetland occurs as a result of the natural variability in soil characteristics but also through the influence of different utilization techniques and the proximity to and influence of water channels, natural vegetation and springs. In addition, micro-variations in the surface topography throughout each wetland may also play a significant role in affecting drainage and runoff relationships.

The inherent variation in the water table characteristics throughout drained wetlands inevitably has some effect on crop production although the extent to which this situation can be improved through hydrological management, particularly through drainage modification, is unclear. Whilst the K_{sat} data suggests that a denser drainage network would be required, there are clearly logistical and economic limits to such an undertaking. An alternative strategy to improve the hydrological situation, however, would be to examine the effects of cultivation or grazing on soil characteristics and subsequently promote soil management practices or systems of crop rotation which reduce nutrient loss and structural degradation.

The following chapter, in part, examines the evidence to suggest that farmers are aware of the interaction between water, soil and vegetation in their wetlands and that wetland management practices based upon this knowledge are carried out accordingly towards the goal of sustainable utilization.

Chapter 8

Indigenous Wetland Management in Illubabor

Introduction

This chapter focuses on the perspective of those communities who regularly interact with wetlands in Illubabor zone, drawing upon information collected during a programme of PRA sessions described in Chapter Five. Farmers' perceptions, understanding of wetland functioning and their utilization strategies are discussed, thus establishing the characteristics of the human – wetland interactive process. More fundamentally, the evidence to suggest that local people's knowledge of wetlands bears the typical characteristics of an indigenous knowledge system, as highlighted in Chapter Three, is examined. This has important implications for the ability of communities to adapt to changing environmental and socio-economic circumstances and, ultimately, the sustainability of the wetland system.

The data

The data are the result of a programme of PRA sessions which took place at Dizi, Bake Chora, Tulube, Anger and Supe wetlands between February and May 1998. Supe, which was not included in the initial hydrological classification, was included in the PRA programme primarily as a result of other on-going research activities being carried out under the co-ordination of the Ethiopian Wetlands Research Programme (EWRP). These included both hydrological investigations and a PRA programme which were initiated during 1998. Although the PRA data from Supe cannot be related directly to any hydrological data as a result of an incomplete hydrological data set, it is nonetheless contextual to the overall study and, in some cases, a means of triangulating research findings. Beyond the site specific level, data from Supe wetland are also useful in contributing to the wider debate on the mechanisms of adaptation and evolution of IK.

The lack of current human intervention in their management meant that the sites of Chebere and Hurumu were not included in the PRA programme. In addition, although PRA sessions were initiated at Wangeneye wetland, communication problems were experienced between researchers and farmers, and the site was subsequently excluded from the PRA programme.

Farmers' perceptions of wetland hydrology

At each of the wetlands farmers played an active role in producing seasonal diagrams of hydrological variables which included rainfall, water table and discharge from the wetland outflow. In addition, many farmers were able to produce seasonal diagrams of water table variation in both cultivated and uncultivated areas of wetland. During activities such as transect walks and group discussions, farmers demonstrated knowledge of intra-site water table variation, water colour changes throughout the year and factors which they considered significant in affecting wetland hydrology. Examples of the latter include perceived variation in soil characteristics throughout the wetland, the location of springs and the influence of the depth of soil to the bedrock.

In many cases farmers also exhibited extensive knowledge of changes in the hydrology of their wetlands over time. This section presents much of the information generated throughout the PRA sessions, drawing attention to the characteristics and variation in farmers' hydrological knowledge.

The farmers' perceptions of seasonal rainfall

The patterns of rainfall highlighted in the seasonal diagrams are the result of farmers being asked to describe the typical rainfall pattern in their particular wetland throughout the year. The general pattern of rainfall was unanimously recognized as one of a peak around July and August, with a decline to the driest period in December and January, confirming the expected seasonal pattern of rainfall (Figure 4.1). There are, however, some key differences in the farmers' perceptions of monthly rainfall between sites (Figure 8.1). In particular:

There are relative differences in perceived rainfall levels between each successive month. For example, at Anger wetland, the difference between July and August is greater than the difference between the same months in Dizi wetland.

There is variation in the observed highest and lowest months in the rainfall calendar between sites. For example, in Dizi wetland, January and February represent lowest rainfall months of the year and August the highest. In contrast, January is the lowest and July the highest in Bake Chora wetland. This relationship for all sites is summarized in Table 8.1.

Perception of rainfall levels at the start of the calendar is variable between sites. In particular, between September and November perceived rainfall appears to follow two patterns. The first is typified by Dizi and Anger wetlands where perceived rainfall levels gradually decrease between September and November. In Bake Chora, Supe and Tulube, however, rainfall during October is perceived as less than that of both September and November. Moreover, one group of farmers at Tulube argued that there was a complete absence of rainfall during October.

Indigenous Wetland Management in Illubabor 151

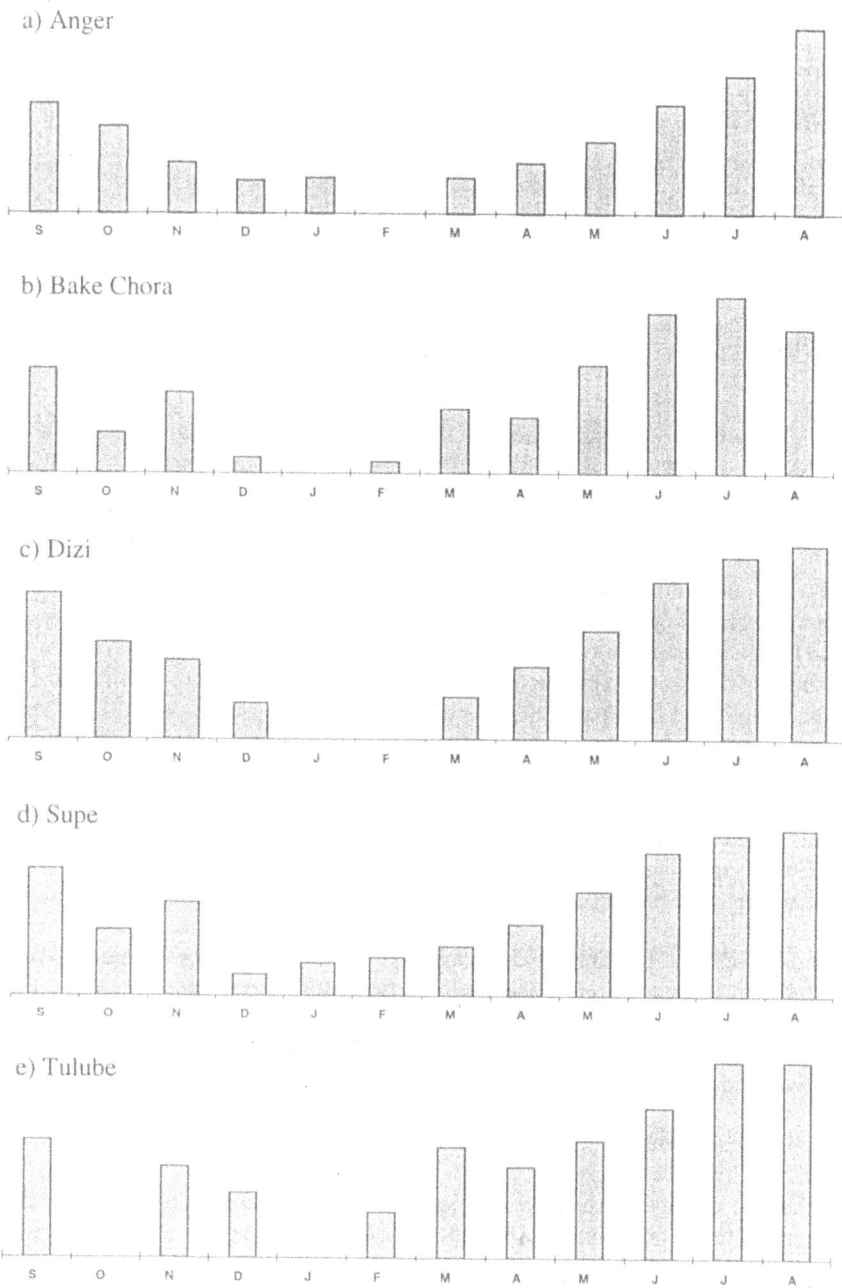

Figure 8.1 Farmer perceptions of rainfall at each wetland

Table 8.1 Differences in perceptions of high and low rainfall levels between sites

	Rainfall Lowest Month		Highest Month
Dizi	Jan - Feb	(no rain)	Aug
Tulube session 1	Jan - Feb	(no rain)	Jun
Tulube session 2	Jan, Oct	(no rain)	Jul
Supe	Dec	(some rain)	Aug
Anger	Feb	(no rain)	Aug
Bake Chora	Jan	(no rain)	Jul

As a means of triangulating this information, many farmers were asked to describe how the previous year's rainfall differed from that demonstrated. Tulube's farmers claimed that there was more rainfall towards the end of the previous year (April, May 1997) and, in agreement with Supe's farmers, at the start of the current year (October, November 1997), the effect of which was a noticeable increase in water table elevation throughout this period. In addition to flooding, Supe's farmers suggested that the abnormal rainfall also had a significant effect on the upland crops, in particular causing damage to coffee.

During a PRA session at Anger wetland during May 1998 farmers pointed out that the rainfall had been particularly late in their present calendar and that when it did arrive, the intensity had been much greater than usual. Prior to this, the delayed rains were of particular concern to farmers at Bake Chora, who pointed out that the drought conditions favoured the development of a parasite which feeds on their wetland maize crop.

The seasonal diagrams of rainfall illustrate that farmers have a clear idea of the expected pattern of rainfall throughout the year and that they are perceptive to any deviations from normal conditions. Furthermore, they also demonstrate some recognition of the consequences of abnormal rainfall conditions, particularly its effect on the water table elevation and consequently crop production. The PRA sessions have, however, revealed distinct differences between the perceptions of rainfall at different sites, which seem unlikely to correspond to meteorological data, given the relatively close proximity of each wetland to each other.

Farmers' perceptions of water table and runoff relationships

In contrast to rainfall, farmers' perceptions of water table levels are less variable between wetland sites (Figures 8.2 and 8.3). The yearly cycle of a rising water table from May until July or August, followed by a decline to January or February is universally perceived by farmers throughout the study area and this view is compatible with the general trend in water table level throughout the year (Figure 7.7). As in the farmers' perceptions of rainfall, however, there is some variation in perceptions of the highest and lowest months of the year (Table 8.2). Those

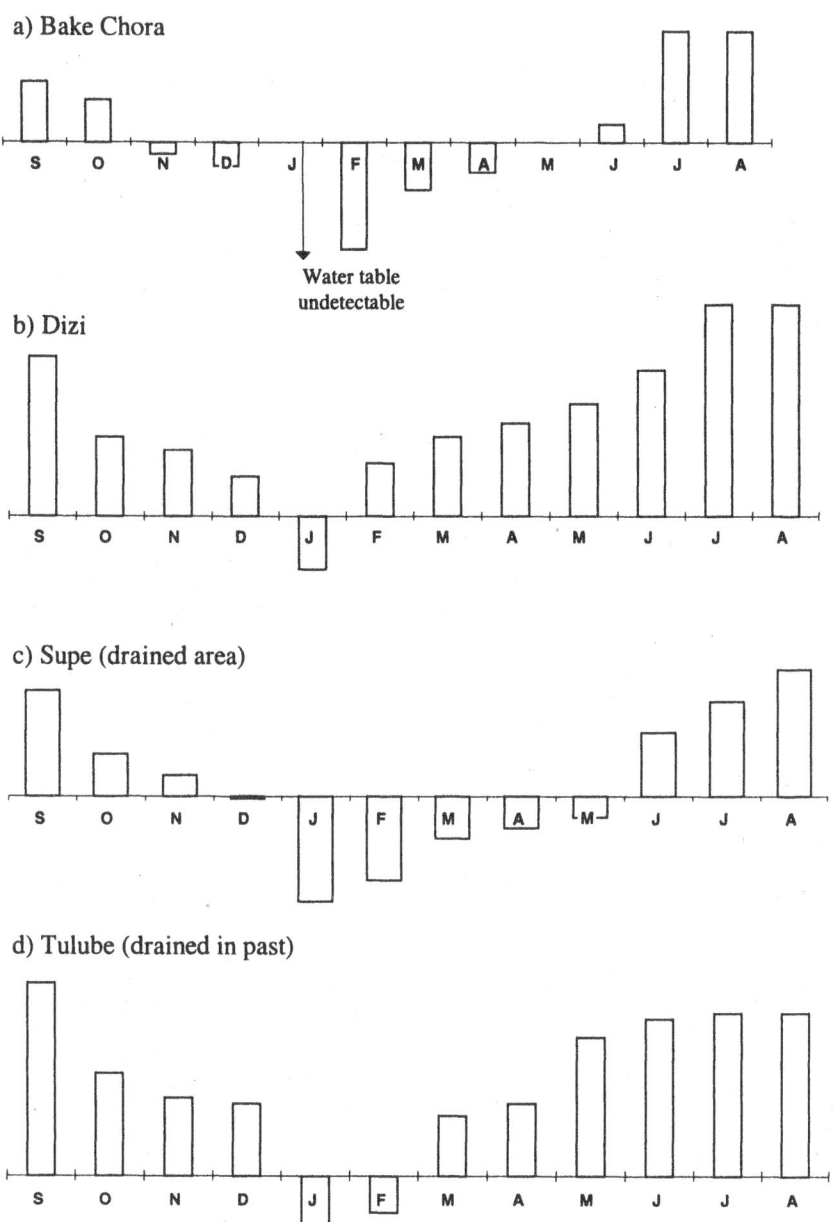

Figure 8.2 Farmer perceptions of water table elevation at each site (cultivated)

a) Tulube

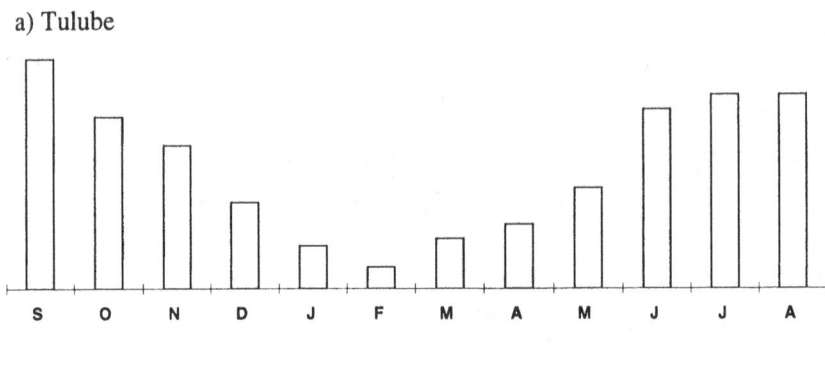

b) Supe (undrained area)

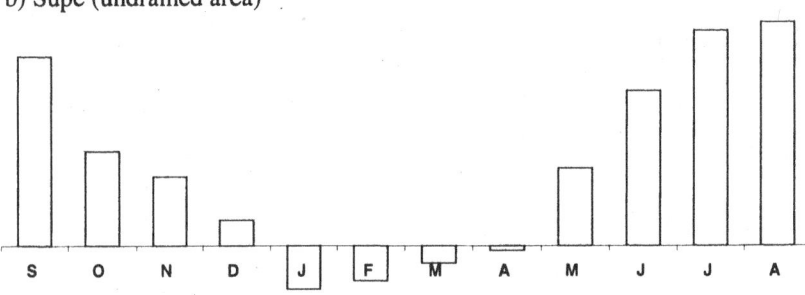

c) Anger

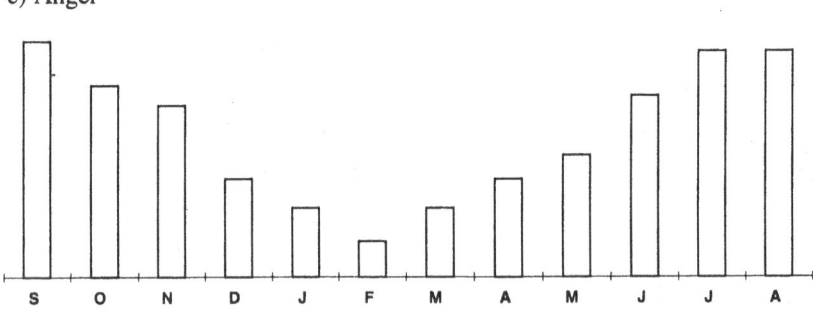

Figure 8.3   Farmer perceptions of water table elevation at each site (uncultivated)

Table 8.2 Differences in farmers' perceptions of high and low water table levels

	Water Table Highest Month	Lowest Month	Land Use
Dizi	Aug	Jan	Crop cultivation
Bake Chora	Aug	Jan	Crop cultivation
Supe (drained area)	Aug	Jan	Crop cultivation
Supe (undrained area)	Aug	Jan	*Cheffe* harvesting
Anger	Sept	Feb	Crops / grazing / *cheffe*
Tulube [1] (present)	Sept	Feb	Grazing
Tulube [2] (present)	Sept	Feb	Grazing
Tulube [1] (past)	Sept	Jan	Crop cultivation

[1] session held on 2 March
[2] session held on 6 April

wetlands with the highest water table level in August (Dizi, Supe and Bake Chora) reach their lowest level in January (Figures 8.2a, 8.2b, 8.2c and 8.3c) and, with the exception of farmers' perceptions of Tulube in its previous drained state, those wetlands with a highest water table in September (Anger and present day Tulube) reach their lowest in February (Figures 8.3c and 8.3a). Closer examination reveals that those wetlands where the August – January 'cycle' is perceived (Dizi, Bake Chora and Supe) are currently undergoing cultivation. Meanwhile, the September – February 'cycle' is perceived in those sites where there is either limited or no cultivation taking place. This suggests that a key impact of drainage in the wetlands is the depletion of the water table to its lowest point approximately one month earlier than undrained wetlands. It is, however, unclear as to why the farmers in these drained wetlands suggest that the water table starts to increase during February whilst on undrained sites it reaches its lowest level during this month.

Whilst the variations in the pattern of perceived water table throughout the year appear to be explained by the presence of drainage, the data from Supe wetland suggests a deviation from this pattern. Here, farmers' perceptions of water levels in the drained area (Figure 8.2c) are essentially similar to those of the undrained area (Figure 8.2b). Whilst this may be the case, it is important to recognize that the data are only representative of those farmers who attended the PRA sessions and that wetland use is spatially variable. At Supe wetland, those farmers attending the PRA sessions acknowledged that they were not sure about water levels in 'natural' areas and in Anger, the farmers who constructed the seasonal diagrams stated that they were not presently engaged in wetland cultivation and were therefore reporting on a non-cultivated situation.

Figures 8.2 and 8.3 also highlight further perceived differences between drained and undrained wetlands. In Figure 8.3 Supe is conspicuous in that water table levels are perceived to be below the wetland surface for four months during the year in contrast to the other undrained sites (Tulube and Anger) where water levels constantly remain above the surface. In addition, perceived water tables in the seasonal diagram of Dizi wetland (Figure 8.2b) are inconsistent with those suggested for other cultivated sites in that the water table is perceived as being above the surface for 11 months of the year. This renders it more similar to the undrained wetlands.

Although the extent to which these perceptions represent hydrological reality is discussed further in Chapter Nine, again it should be reiterated that the spatial variability of the water table in each wetland could offer some explanation for any inconsistent perceptions indicated on the seasonal diagrams. Those farmers attending PRA sessions are likely to convey the knowledge they have of their own wetland plots which undoubtedly have a range of different drainage requirements which may differ significantly from the overall hydrological characteristics of a particular wetland.

Despite the inconsistencies between different wetlands, farmers' perceptions of water table levels during the year generally demonstrate a knowledge of the impact of drainage on these sites. This impact appears to be a lowering of the water table from natural conditions so that the water level remains below the surface of the wetland for a longer period of the year although the perceived duration of this lowering differs between sites.

Following the production of these seasonal diagrams, several groups of farmers elaborated on the relationship between rainfall and water table levels in their wetlands. When asked if there was any time period separating an increase in rainfall and water table farmers at Bake Chora proposed that during the dry season, rainfall has no effect on the wetland water level. During the wet season they suggest that water level starts to increase within one hour of rainfall:

> ...if it has been raining continuously and the upland becomes full, then within one hour the water level starts to increase. The water comes from every corner but mostly from the springs (farmer at Bake Chora, 14 April 1998).

This suggests some knowledge of the storage capacity of the catchment in that the impact of rainfall during the dry season is unlikely to affect the flow of water from springs. During periods of continuous rainfall, however, the saturation of the soil promotes surface and sub-surface flows from the catchment to the perimeter springs, hence the observed rise in wetland water table levels. At Tulube farmers suggested that it must rain for several weeks before water starts to flow across the surface of the wetland, because the wetland soil must first become fully saturated.

A comparison of rainfall and water levels in Supe wetland (Figures 8.1d and 8.2c) reveals that as rainfall begins to increase in January the water level reaches its minimum levels. As in Tulube and Bake Chora, farmers at Supe offer the explanation that the rainfall is initially stored in the upland area:

> In December, January and February there's not enough rain to run into the wetland and it stays in the upland area (farmer at Supe, 15 April 1998).

These explanations go some way in accounting for the fact that farmers generally represented the water table in their wetland as less variable than they did the rainfall, i.e. water levels during the year follow a pattern which is only seriously influenced by rainfall later in the season. The implication is that, if these graphs are reasonably accurate, the catchment performs a buffering role which tends to filter out the rainfall variations shown in Figure 8.1 and consequently produces the common seasonal water table patterns illustrated in Figures 8.2 and 8.3.

Clearly the farmers have some knowledge of the inter-relationship between rainfall, water levels and, significantly, the storage capacity of their catchment and wetland. There is also consensus that springs are the major source of water in the headwater wetlands (Bake Chora, Tulube and Supe) whilst also playing a significant role in contributing to the main streamflow in the mid-valley wetlands. Apart from contributing to the overall hydrological regime, they are also valuable in the supply of potable water, which according to farmers, is not collected from any other source.

In view of their dependence on springs, it is of little surprise that most farmers have an in-depth knowledge of the changes in water colour which occur throughout the year. Table 8.3 shows the changes in water colour observed by Tulube's farmers and similar evidence is presented by Supe's farmers, who also suggested there are health problems associated with discoloration. Farmers in Anger and Dizi also understand discoloration to be an indicator of upstream soil erosion. In contrast, farmers at Bake Chora maintained there were no such changes in water colour during the year, although the farmers construct a fence around their main headwater spring to prevent contamination from farm animals.

Table 8.3 Tulube farmers' perceptions of water colour

	Location in wetland		
	Head	Middle	Mouth
Dry Season	'clean'	'reddish colour'	'no water'
Wet Season	'white colour'	'cloudy'	'deep red cloudy'

A unifying theme of the PRA sessions which centred around wetland hydrology and, to some extent those sessions which did not, was that of the spatial variability in wetland characteristics, in particular, the wetland water table. The seasonal diagrams constructed by farmers represent a generalized account of the wetland water table in each wetland by their own admission. In reality, farmers acknowledged that water levels were variable throughout each wetland and several reasons for this variation were offered, the most frequent being the influence of rocks, which was summarized by one Supe farmer:

> The lower part of the wetland is always wetter. When we try to drain this area we can't dig it properly because of the stones near the surface. The stones block the water and keep it in. When there's excess rainfall, the bottom becomes wet first and the moisture

spreads upwards towards the head of the wetland (farmer at Supe, 15 April 1998).

In addition, rocks are seen as a major impediment to soil drainage at Dizi wetland resulting in a damaged maize crop:

> ...some areas have a level surface and underneath these areas are stones. These stones block the water and cause the level of water to increase. These areas are therefore much wetter than uneven areas and the crops do not grow as well here (farmer at Dizi, 26 April 1998).

At Bake Chora and Tulube, farmers associate the occurrence of rocks with a decrease in topsoil and, therefore, low soil fertility, and farmers at Dizi also recognize that the slope of the wetland surface affects the level of 'underground water'. One farmer described how the sides of his wetland area were relatively steep and how water did not stay in that area. Instead it collected in a more level area further down the wetland, which was subsequently very difficult to drain. Another Dizi farmer owns a plot of wetland where:

> ...the soil here is very sandy and it will not hold any water. Even if this area is irrigated by hand it becomes dry very quickly (farmer at Dizi, 26th April 1998).

Finally, a lack of or inefficient artificial drainage is regarded as a cause of waterlogging in some areas of cultivated wetlands. For example, in Bake Chora some wetland plots have been abandoned for unknown reasons and other farmers are becoming concerned about the effect of these areas on their own land, in particular the effects of inadequate drainage on crop growth. Furthermore, farmers agree that waterlogging is exacerbated by the growth of natural *cheffe* vegetation which is regarded as a weed by some but a valuable resource by others.

Not only were farmers able to comment on the hydrological regime in their own individual plots they were also to some extent, aware of what was going on in their fellow farmers' plots. Such knowledge of hydrological variation and how factors such as soil type and rocks are indicative of soil moisture conditions, appears to be fundamental to the design and implementation of their drainage networks. Furthermore, this knowledge interacts with that of the specific water requirements of different crops, leading to a strategically designed drainage system. The only constraints in this system of hydrological control would appear to be those which are beyond the farmers' control and essentially socio-economic in nature.

Wetland utilization

Farmers' knowledge of wetland hydrological processes has been acquired largely through their interaction with the wetland resource which has taken a variety of different forms. The PRA programme, however, identified five forms of wetland utilization, as described by farmers. These included:

1. the cultivation of crops following drainage of the wetland;
2. the harvesting of natural *cheffe* vegetation for a variety of uses;
3. the use of wetlands as grazing areas for farm animals;
4. the provision of a clean water supply;
5. as a source of medicinal plants.

Wetlands are not limited, however, to one particular type of use. The different uses listed above represent the building blocks of an overall wetland utilization strategy in which each farmer may use the wetland in a variety of ways to meet his specific needs, resulting in a single or multiple use at one point in time. Furthermore, this operates at two levels: the individual farmer's wetland plot and the wetland as a whole. Consequently, the inter-relationship between the different types of utilization is very complex as farmers attempt to manage each of the wetland's resources in a way that retains the maximum utility relative to their needs. Evidence from each wetland suggests that this is not an easy task, as wetland utilization strategies have clearly changed over time in response to a range of pressures. While some wetlands such as Tulube have been sporadically abandoned, others like Bake Chora have reportedly been cultivated continuously for over 80 years.

Wetland cultivation

By far the most conspicuous form of wetland utilization is drainage which provides agricultural land suitable for crop cultivation. One of the assumptions prior to the start of this research was that intensive wetland cultivation was a relatively recent phenomena which came about through the influence of several factors such a population growth, political and economic change. One of the findings of the PRA programme was that wetland cultivation appears to have been carried out over a much longer time (between 40 to 80 years ago) and according to most farmers, it exists today in much the same form as it did in the past.

The principal reason given for draining and cultivating wetlands is the alleviation of food shortages. Farmers at Dizi maintained that there was always a shortage of food during the rains (June and July) and wetland cultivation is seen as a means of bridging this gap until the upland harvest. Together with farmers at Bake Chora, there is consensus that if food shortages did not exist during these months, they would still cultivate their wetlands but for more market-orientated crops such as sugar cane and vegetables, rather than maize.

The principal techniques involved in the drainage and cultivation of wetlands are typified by the seasonal calendar produced by farmers at Supe (Figure 8.4). This highlights the order in which a range of wetland farming practices are carried out during each year, many of which are fundamentally similar to those practices in other cultivated wetlands. These core activities undertaken by farmers include:

1. Drain management, which is carried out at the start of the season and usually involves the clearing of old drainage ditches or the excavation of new ones. This is regarded as an essential part of the wetland farming system as it

regulates the amount of water residing in and flowing out of the wetland. The blocking of ditches is also carried out early in the season as a means of increasing the residence of water behind the blockage which improves the wetland soil. Ditch blocking is also used as a means of regulating soil moisture throughout the wetland (Tegegne Sishaw, 1998) (see below).
2. Weeding (known locally as *jala hara*) is undertaken throughout the farming calendar in an attempt to reduce the competition for moisture and nutrients between the main crop and other invading species. Weeds play an important role as indicators of moisture and soil fertility.
3. Burning involves the collection of the previous year's maize stalks and burning them to increase soil fertility. The ash produced is then ploughed into the soil.
4. Ploughing may be carried out several times within a year although usually immediately before and after the sowing period. In the first instance it is commonly used as a means of loosening the soil structure (a process known as *babaka*) and additionally it can be used after sowing to bury the seeds.
5. Sowing is undertaken by hand using the broadcasting or row planting methods, depending upon the time and labour constraints of the farmer.
6. A second period of weeding, known locally as *charega*, involves the identification of the best individual maize plants in terms of the quality of their crop, followed by the destruction of diseased or stunted plants which are competing for nutrients.
7. Guarding is regarded of primary importance to most farmers. Guarding from pests is carried out from the moment seeds are sown, as pest damage is widely reported as one of the biggest problems facing farmers.

In each wetland these practices appear, however, to be more specifically adapted and constitute a more individual range of activities, many of which are not common to all the wetlands. Table 8.4, which has been compiled from a comparison of the seasonal farming calendars produced from each wetland, shows these more specific activities and the interpretation of the names given for each during discussions.

Although the range of farming activities highlighted in Table 8.4 suggests some differences between sites in terms of the actual activities that are undertaken, it is possible that these differences may be a result of the misinterpretation of farmers' comments during PRA sessions. For example, 'clearing' and 'weeding' are rather ambiguous descriptions which may refer to the same activity or anything ranging from oxen 'weeding' (also a type of ploughing) to the 'clearing' of diseased maize stalks, a practice described in other wetlands as 'thinning' and 'selection'. Another example is whether 'ditch clearing' and 'ditch digging' represent the same activity. Although these could commonly refer to the practice of ditch excavation, it is also possible that these are in fact different practices and, on the basis that farmers are well aware of the hydrological management requirements of their own wetland plots the adoption of specifically adjusted techniques is highly feasible.

Indigenous Wetland Management in Illubabor 161

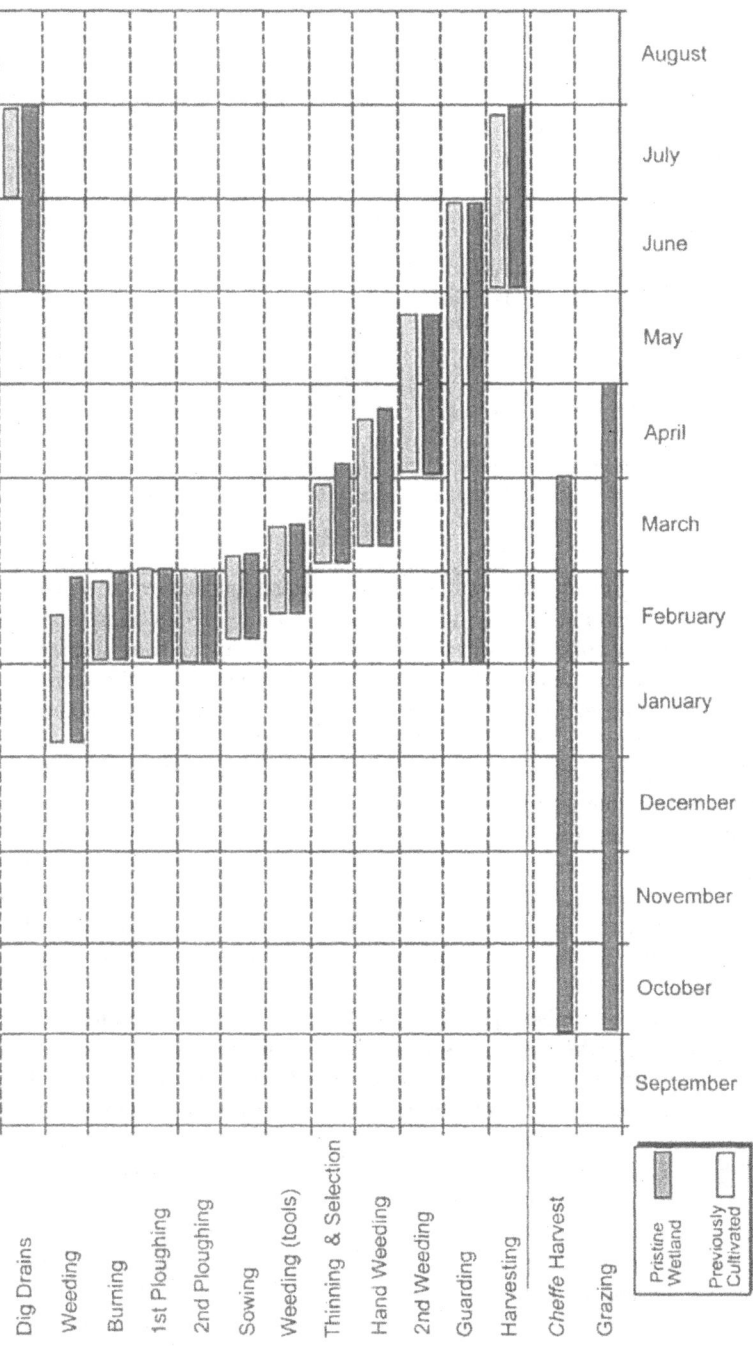

Figure 8.4 The seasonal calendar of wetland farming activities produced by Supe farmers

Table 8.4 Summary of wetland seasonal farming calendars

Activity	Dizi	Supe	Tulube	Bake Chora
'Drain Digging'		Jul		
'Ditch Clearing'	Sep		Sep, May, Jun	Sep/Oct/Dec/Apr
'Ditch Blocking'				Sep
'Grass Clearing'	Jan	Jan/Feb		
'Weed cutting'			Oct	Nov
'Burning'	Jan	Feb	Oct	Nov
'1st Ploughing'	Feb	Feb	Oct	Nov
'2nd Ploughing'	Feb/Mar	Feb	Oct	
'Sowing'	Feb/Mar	Feb/Mar	Dec	Dec
'Hand Weeding'	Apr	Feb/Mar	Jan/Feb	Dec/Jan/Feb
'Thinning/selection'		Mar	Mar/Apr	('head cutting') Apr
'Oxen Weeding'	May		Feb/Mar	
'2nd Hand Weed'	May/Jun	Mar/Apr		
'Clearing'	Jul			
'Harvest'	Jul/Aug	Jun/Jul	Jun - Aug	Jun/Jul
'Guarding'	Feb - Jun	Feb - Jun	Dec - Aug	Jan - Aug

There are some activities, however, which are clearly undertaken at some sites and not others. For example, only the farmers of Bake Chora mentioned the blocking of drains during September as a means of improving the fertility, structure and moisture content of the wetland soil (Figure 8.5) (although ditch blocking was also observed at Wangeneye where PRA sessions were initiated but later abandoned as a result of a lack of participants). The reasons for this are unclear although of the other sites where PRA sessions were undertaken, both Anger and Dizi constitute mid-valley wetlands which would be unsuitable for this practice on the basis that their water supply via the natural stream channel is more consistent throughout the year, unlike the spring fed headwater wetlands. There would, therefore, be less of a need to assist flooding, even in the artificial feeder drains. At Supe, farmers already report problems with waterlogged soils which reduces the need for ditch blocking and at Tulube, drainage is not currently practised.

Bake Chora's farmers also specified the least number of wetland farming activities (ten in total) which suggests that wetland farming activities are either limited by constraints or that the wetland farming system at this site has been refined and activities are more effectively managed. Given that wetland cultivation has been undertaken at Bake Chora for approximately 90 years, which is longer than any other study site, this gives some credence to the latter suggestion.

Further variation in the wetland farming calendars is evident in the timing of farming activities. Analysis of the seasonal calendars reveal that wetland cultivation begins much earlier in Bake Chora and in Tulube's past farming calendar than Supe and Dizi. In the former, drainage begins at the start of the Ethiopian year in September and significantly, on several other occasions later in

Figure 8.5 The practice of ditch blocking as a means of regulating water supply to the wetland (pictured here at Wangeneye)

the year (December and April). This is in contrast to Supe and Dizi where drain maintenance is carried out only once. In addition, key events such as sowing and burning are carried out in Bake Chora approximately two months ahead of their counterparts in Supe and Dizi wetlands.

The most obvious explanation for these differences would be that they are on-site adaptations that have occurred in response to environmental differences (hydrological) but also within the socio-economic context of the wetland's community (e.g. good organization and co-operation). Alternatively, they may also reflect socio-economic constraints, in for example, the upland farming system which may divert resources away from the wetland cultivation system at different times of the year.

In terms of the knowledge which is applied in each farming activity, the wetlands also show a broad similarity in the techniques used. The process of drainage and drain maintenance is common to most sites with the only major difference being the chosen location for drains which, according to farmers, tend to reflect the spatial pattern of soil moisture conditions within each wetland. In addition to direct observation, the growth of specific weeds and the yellowing of maize are frequently used as indicators of soil moisture conditions which determine the location of drainage channels.

Drainage tends to follow the method and pattern described by several studies of indigenous wetland use. In particular, Zimmerer (1991) reports that wetland agriculture in the Andes is facilitated by the excavation of a central channel from the outflow of the wetland, up towards the head and this is supplemented with lateral drains where needed. At Tulube wetland farmers described a similar process:

> When wetland drainage first started, we all discussed how it should be carried out and then decided that it should begin with the excavation of a central drainage channel which follows the natural watercourse. Several drainage ditches were then dug at the side of this channel depending upon where there was excess soil moisture, which is indicated by a change in the colour of the maize crop from green to reddish (farmer at Tulube, 24 February 1998).

The depths of drainage channels throughout the study wetlands are altered according to the moisture conditions at that specific time of the year and these are reported as varying between 50 cm reaching a maximum of 1m. No farmers reported making any modifications to the width of drainage channels, which tend to remain approximately 50 cm. The reasons for this are unclear although it is likely that any increase in the width of ditches would reduce the area of cultivable land throughout the whole wetland. A further requirement for an efficient drainage system is constant maintenance, in order to reduce the effects of weed infestation or the build up of sediment which can itself be a precursor to weed growth.

A wide variety of tools are utilized throughout the wetland farming calendar ranging from hoes which are used to clear ditches to spears which are used to scare animal pests. A detailed analysis of tool utilization was carried out at Tulube and Anger wetlands where in total, twelve different tools were specified. Table 8.5 shows the range of tools identified at these wetlands. Noticeably, farmers appear to employ a range of different tools to perform tasks which seem very similar. For example, *Gasso* and *Sapata* are both used for digging ditches, while *Buto*, *Koto* and *Haamtu* are all used for cutting vegetation in the wetland. The fact that these were all specified by farmers implies that they do perform different tasks and may represent specialized adaptations which may have been lost in the Oromiffa – English translation.

Table 8.5 Tools utilized in the wetland farming system

Oromiffa	English	Use	Tulube	Anger	Supe
Gasso	hoe	Digging ditches / weeding / soil dispersal	•	•	•
Togo	machete	Cutting grass and shrub roots	•		•
Sapata	unknown	Digging ditches / weeding / soil dispersal	•		
Hoko	rake	Raking vegetation	•		•
Buto	sickle	Cutting reeds, weeds and grass	•		
Kotto	axe	Cutting vegetation	•		
Wanchife	sling	Throwing small stones at pests	•		
Aybo	spear	Scaring animal pests	•		
Gajeera	knife	Cutting the drains / pest scaring		•	
Akafa	shovel	Clearing grass from ditches		•	•
Haamtu	sickle	Cutting the roots of weeds	•	•	
Faci	axe	Cutting tree roots		•	

Wetlands as a grazing resource

In addition to the cultivation of wetlands farmers also utilize their wetlands as grazing resources, usually on patches of wetland where cultivation is not taking place or after the harvest. Of the study wetlands where PRA was undertaken, grazing was identified as a major wetland use at Tulube, which is regenerating following cultivation, while at Bake Chora and Supe animals have restricted access because of on-going cultivation. At Supe, cattle are grazed only in the uncultivated area of the wetland and only during the dry season. In Bake Chora:

> During the dry season they [the cattle] prefer the wetland. In the wet season they prefer upslope. In the wetland they eat *komate*, weeds and *cheffe*. There is not enough grass in the wetland but they still prefer it during the dry season. During harvest time the cattle are left to wander where they like (farmer at Bake Chora, 10 March 1998).

Farmers at Supe did recognize, however, that cattle produced certain problems in the wetland, in particular the destruction of the drainage network and the compaction of the wetland soil. At other sites such as Anger and Hurumu, those areas of wetland which show signs of degradation (low water table and grassy vegetation) were commonly utilized for cattle grazing. If the grazing of cattle in these wetlands is a destructive process, this is of particular concern given that wetlands are increasingly being regarded as important grazing resources, as traditional grazing areas in the uplands are either cultivated or used for coffee cultivation (Afework Hailu, 1998).

Collection of cheffe and natural vegetation

Cheffe (Cyperus latifolius), which constitutes the natural climax vegetation in these wetlands, is regarded as both a nuisance to wetland cultivation and a valuable natural resource. Not surprisingly, most wetland farmers have an intimate knowledge of the role of *cheffe* which has been attained through repeated familiarity during successive seasons of wetland use, whether this has involved cultivation or not.

The harvesting of *cheffe* as part of a farmer's wetland utilization strategy appears to be a case of 'all or nothing', in that farmers either drain their individual wetland plot for agriculture or they abandon it for *cheffe* growth. Multiple use of an individual farmer's plot was observed only once, at Bake Chora wetland, where the farmer retained a small area of *cheffe* at the side of his cultivated plot (Figure 8.6). This area supplies him with a small quantity of reeds for the roof of his *tukul* but in addition:

> The reeds maintain a supply of water. If the area is drained, the land below will become dry and there will be no underground water (farmer at Bake Chora, 10 March 1998).

This suggests that *cheffe* is understood to perform a role in hydrological regulation (which is reflected in the regulations of the NDPC (Table 4.3), although

Figure 8.6 The head of Bake Chora wetland showing a range of land uses including the reservation of *cheffe*

no other farmer within the study mentioned such a function. By its very occurrence in pristine wetlands, however, it has an association with saturated conditions and several farmers offered explanations of this relationship, based upon their experiences of attempting to eradicate *cheffe* during the wetland cultivation process. The major problem is that if drainage is not efficient, the *cheffe* will inevitably start to recolonize as a result of its dense network of rhizomes below the surface. Furthermore, as the *cheffe* recolonizes it promotes widespread waterlogging which enhances further growth. As portrayed by the farmers, this whole system of *cheffe* recolonization appears to be a relatively rapid process which needs to be kept in check if hydrological management and cultivation is to be successful. This situation, however, only appears to be the case where wetland cultivation continues to be carried out annually. On those sites where complete or partial abandonment has occurred following several years of cultivation, farmers suggest there is an intermediary phase which precedes complete *cheffe* recolonization. For example, at Tulube farmers suggest that where the soil is deep the plant *tufo* (*Asteraceae*) germinates first and this is then followed by *cheffe*. On shallow soils *komate* (*Leersia hexandra*) is the precursor to *cheffe*. According to farmers both *tufo* and *komate* grow roots which penetrate the wetland soil thereby increasing the pore spaces, soil moisture residence and consequently the height of the water table. This in turn is said to promote the growth of *cheffe*.

The implication is that *cheffe* prefers and promotes waterlogged conditions, something the farmer at Bake Chora obviously recognizes and uses to his advantage in providing a water store during the dry season. Although this is the only example of this practice in this study, Afework Hailu (1998) reports that elsewhere in Illubabor wetland farming communities operate a system which involves reserving the head of each wetland as a source of *cheffe* to maintain a water supply to the rest of the wetland during the dry season. This is however, part of an overall wetland utilization strategy which involves the reservation of larger areas of wetland rather than areas within individual plots.

Although the formal practice of *cheffe* reservation was not observed in the study wetlands, plots of *cheffe* do exist where farmers have abandoned cultivation on their individual plots for a variety of reasons. An example of this situation is Anger wetland (Figure 8.7) in which some farmers prefer to grow *cheffe* in their plots while others cultivate maize. Those who reserve their land for *cheffe* do so because in the past they suffered from a shortage of roofing material. Consequently they value *cheffe* more than the equivalent output from the use of the land for cultivation. The effect for the wetland as a whole is a 'patchwork' of land use and, as suggested by the farmers at Anger, a readily available supply of *cheffe* for the wetland farming community.

Figure 8.7 *Cheffe* **reservation alongside maize cultivation in Anger wetland**

To a lesser extent, this situation is also represented by Bake Chora wetland where several farmers in the middle and the bottom of the wetland do not cultivate their land and instead use the *cheffe* from these areas for the roofs of *godo*, the small shelters used during pest guarding. Here the owners of these areas are utilizing the *cheffe* but there is no evidence either way to suggest that these form part of a conscious attempt at hydrological regulation (or indeed *cheffe* cultivation itself, which may be a by-product of abandonment caused by other constraints and pressures). Nonetheless, such variable patterns of land use within each wetland will inevitably have some influence on the wetland hydrological regime and the extent to which farmers' hydrological management practices are effective.

In addition to the intended reservation of *cheffe* and its use as a by-product of abandonment, the utilization of *cheffe* is also a by-product of a strategy of fallowing linked to the cultivation system. In Dizi, for example, one farmer described how he left an area within his wetland plot fallow:

Before one year, this area was being cultivated and then it was left fallow. The cheffe has increased, along with 'inchinne' and the moisture level. Ten years ago this area gave a good yield of sugar cane but the problem with sugar cane is that it rapidly exhausts the soil fertility. When the maize was planted here it became very weak so it was left fallow to allow the fertility to recover. This year the vegetation will be cleared and sugar cane will be planted again (farmer at Dizi, 26 April 1998).

This example highlights the importance of other wetland plant species which, according to farmers, play an important role as indicators of soil fertility and moisture. The plant known locally as *inchinne* is regarded as an indicator of good soil fertility whereas *komate* is found on poor soil. Dizi farmers suggest that in the past when *komate* became abundant, they blocked their drainage ditches until *cheffe* began to recolonize.

Knowledge of other plant species also appears to be extensive. Kumelachew Yeshitela (1998) reports that during a transect walk with farmers at Tulube wetland, a range of plant species were identified which could be used for medicinal purposes, including *balawarante* (*Hygrophila auriculata*) – a skin disease treatment – and *busuke* (unknown) which was used as an enrichment in children's food. In contrast, farmers at Anger claimed that *cheffe* was the only plant they collected from the wetland and at other sites, no farmers offered information about the collection of other plants. It would appear unlikely that the collection of plants is this limited, especially in view of the fact that traditional medicine is well developed in the rural highland societies of Amhara, Tigray and Oromo (Abbink, 1995).

What is clear from the range of discussions held with farmers is the great importance they place on *cheffe*, irrespective of the problems it causes during wetland cultivation. The scale of its utilization for local housing should not be underestimated and, as shown in the case of Anger wetland, its value as a resource is often greater than the potential value of converting the equivalent area of wetland to productive farmland. Afework Hailu and Abbot (1999) also report that this was the case in Chebere wetland, which was abandoned in 1991 when farmers considered *cheffe* more important than their wetland crop. At the same time, the highly prized *cheffe* may also have been the key to hydrological stability, playing a critical role in creating conditions which help maintain and release a supply of water throughout the year. The extent to which all wetland farmers are aware of such a regenerative function, however, remains ambiguous, as evidence suggests that the abandonment of wetland areas has more to do with farming constraints than the search for wetland regeneration.

The constraints on wetland management

The problem of crop destruction by wild pests is almost universally acknowledged by farmers to be one of the biggest constraints on the wetland farming system at the present time. This requires a significant input of labour throughout most of the year. Animals such as baboon, vervet monkey, porcupine and wild pig are regarded as the biggest problems. According to farmers, wild pests have always had an impact on the wetland farming system but in recent years this has increased significantly for two reasons. First, the introduction of the prohibition of firearms during the Derg government prevented farmers from killing or scaring wild pests and secondly, the current government has exacerbated the problem by passing laws making it illegal to kill protected wild animal species.

The villagization programme during the Derg regime is also regarded by many as adding to the problem. During this time many farmsteads were abandoned as people were moved into the new villages and this provided an opportunity for animals to breed and increase in numbers. Major forest clearance has also effectively destroyed the natural habitat of many animals, which have subsequently sought refuge on agricultural land and the small areas of coffee forest which surround many wetlands. Pest problems are also worsened as a result of labour shortages caused by increased urban migration and school attendance, and the subsequent inability of farmers to deal with these shortages. At the same time, farmers claim that the pests are becoming more aggressive and less afraid of humans. Several groups of farmers cite the case of the Colobus monkey which has never posed a problem to humans until fairly recently. In the last few years, however, they claim that this animal has learned from the Vervet monkey and is now attacking crops and on some occasions humans.

A second commonly acknowledged constraint on wetland utilization is a shortage of labour and equipment. In particular, farmers mention a shortage of oxen as a major constraint to wetland cultivation, which is largely a result of livestock diseases that have had serious economic consequences. This has had a wide reaching effect on wetland farming, particularly in those wetlands where farmers rely on the loan of equipment and labour from their fellow farmers.

Shortages of labour have also come about for a number of reasons although mainly as a result of young farmers migrating to the towns where they pursue business interests. There appears to be little incentive to remain as part of a wetland farming community because the returns to labour are very low, a fact which is reinforced by the presence of older farmers whose perceived lifestyles have not changed for the better. There has also been an increase in the number of children and young farmers attending school and this has significantly reduced the labour input. As a consequence, those activities which rely on children, primarily pest guarding, have been neglected.

A decline in the fertility of the wetland soils is regarded as less of a constraint to wetland use although it is common to all the study wetlands. Remarkably, there is no evidence to suggest that wetland cultivation in the PRA study sites has ceased in response to a decline in fertility and crop yields. Instead, other problems such as pests, labour and *cheffe* demand were cited as reasons for abandonment. Farmers' suggestions as to why soil fertility in their wetlands has declined included:

1. the accumulating effects of soil erosion from the wetlands;
2. a reduction in the amount of animal manure present in the wetland as a result of animal shortages;
3. plant residues, which are a means of maintaining wetland soil fertility, being eaten by wild pests.

Finally, it is important to reiterate the point made earlier that waterlogging of the soils within these wetlands is a major constraint to their utilization. The presence of excess amounts of water which limits crop growth is also beyond the control of farmers in many areas, despite their best efforts at adapting and modifying drainage strategies. In the case of Dizi, the geomorphological features of the wetland such as its slope, rocks and shallow soil are the major limiting factors to crop production. In other areas, such as Bake Chora, there is a clear need to balance the demands of a water table which is suitable for crop production against the need to induce flooding for the purpose of soil fertility restoration and *cheffe* regeneration. This universal drainage problem (which may be influenced by the sporadic occurrence of *cheffe* in these wetlands), however, may be the one key factor which sustains the wetlands as key agricultural resources from year to year. In other words, if this problem of under-drainage were to be solved, then the knock on effects for fertility, soil structure and water supply may well result in the degradation of the wetland resource with a significant decline in all of the wetland's functions.

Co-operation between wetland farmers

In attempting to deal with the range of wetland utilization constraints, farmers have responded by seeking a greater level of co-operation with their fellow farmers. This can be viewed as an adaptive strategy which increases the chances of successful wetland utilization at both the community and household level, through pooling resources such as labour and, more significantly, wetland management knowledge.

Farmers unanimously agreed that the system of drainage and cultivation is a labour intensive process which requires the co-operation of all the wetland stakeholders, especially when pristine wetland areas are first cleared for cultivation. Drainage, in order to be successful, must be carried out in the context of the hydrological regime of the whole wetland and farmers are aware that they need assistance from their neighbours. Each farmer's wetland utilization strategy is, therefore, dependant on the relationship he has with other wetland farmers, in particular, the level of co-operation between farmers and consequently the integration of the wetland farming community into a cohesive unit, working together within an understanding of each other's wetland utilization goals.

Evidence suggests, however, that there are different levels of co-operation and farmer integration between sites. Bake Chora is a good example of where farmer integration appears to be high. Here, farmers maintain that they assist each other with several activities in the farming calendar. There is an established system of joint responsibility for drain maintenance, farm animals are loaned out even though there is a serious shortage and sharecropping also takes place. The advantages of this level of co-operation appear to be an annual crop which is almost guaranteed to most farmers and a share for those who assist in its production. Under this situation, most of the farmers at Bake Chora are relatively enthusiastic about wetland cultivation, despite the ubiquitous problems of labour and cattle shortages, pests and other constraints. The farmers also argue that if there were enough food on the valley sides to sustain them during the hungry season, they would cultivate potato and sugar cane in the wetland to generate additional income.

In contrast, farmers at other sites complain of a lack of farmer integration at present (Tulube) and in the past (Dizi) which was enhanced by problems such as pests, labour shortages and declining productivity. This led them to adopt alternative farming strategies on the uplands and wetland cultivation ceased or was scaled down. At Tulube, the farmers have become more and more disinterested in farming their wetland as their individual upland farming systems have evolved in relative isolation from other farmers. Consequently, co-operation between farmers is regarded as being at an all time low.

Farmers at Dizi suggested that co-operation between wetland farmers can be a problem when they each pursue different food security strategies. In particular, some farmers prefer to cultivate coffee for the market or sell their labour rather than cultivate wetlands. At present, however, Dizi's farmers are cultivating their wetland and there would appear to be no serious problems in co-ordinating wetland farming activities.

The degree of co-operation between farmers engaging in wetland drainage and cultivation is, in many respects, a function of the balance between the socio-economic pressures at the household level and the potential problems of wetland cultivation. On the evidence of the farmers at Dizi, co-operation also appears to be influenced by the prevailing political situation which can itself become a constraint to wetland use. A shortage of food is the commonly cited reason for farming wetlands and evidence suggests that in the early years of cultivation, co-operation between farmers is usually high as each of them share this common goal. Once problems such as pests, labour shortages and hydrological instability reach a critical level or they occur at a rate at which farmers are unable to adapt their overall strategy, wetland farming becomes a less economically attractive prospect. At this point, farmers face a decision of whether to carry on farming the wetland in the face of these problems or whether to adopt other strategies. The cohesion of the wetland farming system can be strengthened if farmers choose to carry on farming, as undoubtedly part of this process will involve formulating adaptive strategies to the new set of environmental or socio-economic conditions. If, however, some farmers cease their wetland activities, wetland cultivation becomes progressively harder for those who remain until the point where they can no longer cope and the wetland is abandoned.

This may be avoided if the community operates a policy of re-allocating fallow wetland plots, as was the case in many wetlands during the Derg government (Afework Hailu, 1998). Indeed, the lack of such policies at the present is of concern to Dizi's farmers, who argue that their absence has contributed to a lack of wetland farming integration. In Bake Chora, farmers share similar anxieties:

> There are only two farmers who are not cultivating this wetland. Some farmers are cultivating but not maintaining it. If they don't cultivate it then they will have their area taken off them (farmer at Bake Chora, 26 May 1998).

The major implication is that farmer co-operation and full wetland cultivation are regarded as mutually supportive goals by farmers. Additionally, in a wider context, farmer co-operation can act as an initial buffer against increasing constraints driven by environmental and socio-economic forces. When pressures on the farming system exceeds the ability of local communities to cope there may be a shift from sustainable management which fulfils the needs of the farming community, to unsustainable management where the resource no longer fulfils these needs and environmental degradation may occur. Where farmer co-operation and integration is at its highest, however, farmers arguably have a greater capacity to assist in each other's problems and make specific adaptations by combining their IK resources.

Communication and innovation of wetland knowledge

Chapter Three highlighted several examples of research which indicated that farmers engage in an on-going process of experimentation in response to their changing environment and in the face of numerous constraints, which lower their capacity to farm certain areas effectively (Richards, 1985; Chambers *et al.*, 1989, Haverkort *et al.*, 1999). In particular, there is little evidence to suggest that in such a situation farmers continue as passive actors whose IK remains stagnant. Even when a rapid change in circumstances occurs over a short period of time, farmers will still make some attempt to adapt to their new situation. It is, therefore, highly probable that given the wide variety of problems they experience during wetland utilization, Illubabor's wetland farmers are undoubtedly making adaptations to their changing circumstances.

In contrast to these expectations, however, the evidence from the PRA sessions suggest that adaptations are taking place in a variable and sketchy manner. Whilst some farmers are keen to draw attention to the ways in which they have refined their wetland farming system, either through their own efforts or with the help and assistance of others, other farmers remain adamant that their wetland knowledge has been stagnant since utilization began.

The overall pattern which emerges is one of a move from a wetland farming system forced on tenant farmers with little experience to one where farmers have an understanding of wetland hydrology, the techniques and practices to harness this to meet a variety of goals, and knowledge of the implications of wetland use on their livelihoods. Farmers stated that this shift has come about through the acquisition of new knowledge from a variety of sources which includes their fellow wetland farmers, the *kebele*, the government and NGOs and finally, their own innovation.

Farmer to farmer transfer of knowledge

The PRA sessions revealed some variability in the degree to which different farmers from different wetlands communicate (or acknowledged the communication of) wetland knowledge. At Tulube, farmers stated that they had no relationship with farmers from other areas and at Supe and Bake Chora farmers revealed that they frequently engaged in discussions of wetland use at *kebele* meetings and within the communal labour groups of *dado* and *debo*. Dizi farmers, in contrast, were able to represent the extent of wetland knowledge which originated from neighbouring wetland farming communities. Figure 8.8 presents a Venn diagram constructed by farmers during a PRA session at Dizi which addressed the question of knowledge acquisition. As explained by the farmers, the diagram indicates their present day sources of knowledge and the relative influence of each source on their own IK. The knowledge used from neighbouring farmers (3), was specified as information on drainage ditches and the location of springs.

Whilst Supe farmers did not specify their neighbouring farmers as sources of wetland knowledge, they did acknowledge that the 'administration' had an influence on the way they use their wetland. In particular, the *kebele* was described as playing a role in managing disagreements and allocating land. Undoubtedly for most wetland farming communities, these ubiquitous *kebele* meetings and even market day discussions present significant opportunities to discuss wetland farming practices and facilitate the transfer of knowledge.

In contrast to this 'inter-wetland' exchange of knowledge, farmers tended to be much more open about the 'intra-wetland' transfer of information between themselves, i.e. within their own wetland farming community. Co-operation in farming activities is regarded in many instances as critical to the success of individual wetland plots and the sustainability of the wetland as a whole. An example is provided by the farmers in Bake Chora who, during a transect walk, demonstrated an awareness of their fellow farmers' intentions, practices (including errors) and the consequences of these on other wetland plots:

> Here there is too much water in the soil. He (the absent farmer) has to drain it properly with a big ditch. He also has to block the drains in the dry season so that moisture is retained in the soil. The farmer did not maintain this drain properly. We told him to drain but he has left it too late (farmer at Bake Chora, 26 May 1998).

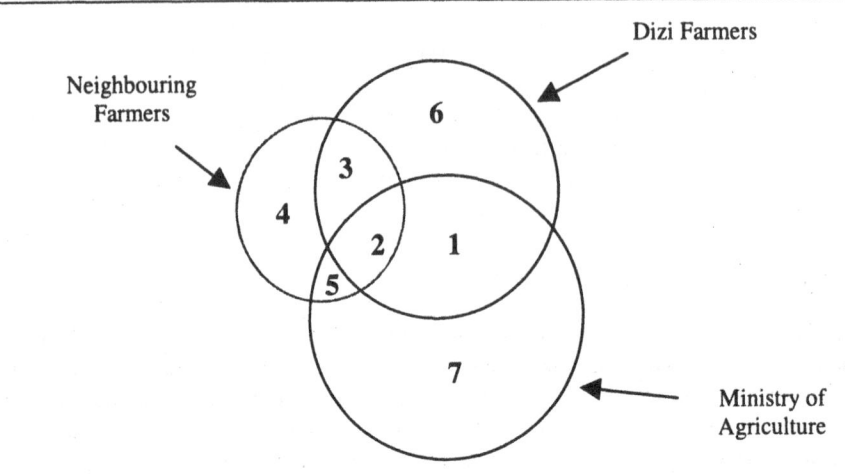

1. The wetland information which originates from the Ministry of Agriculture and which they adopt.
2. The overlap between other farmers' ideas and the MoA, which the farmers are using.
3. The knowledge and ideas exclusively from other farmers which are being used.
4. Unused wetland knowledge from other farmers.
5. The overlap between other farmers' ideas and the MoA, which the farmers are not using.
6. The farmers own exclusive pool of wetland knowledge.
7. The wetland information from the MoA which is not adopted.

Figure 8.8 Dizi farmers' wetland management knowledge and its origins

Tulube's farmers, whilst stating that they never talked to farmers from other wetlands about wetland management, acknowledged frequent discussion amongst themselves:

> We talk together [about wetlands] because if your neighbour can't cultivate their wetland plot, then there will be a problem with your yield. The problem is that if your neighbour doesn't drain, there will be a build up of water on your land so it will be impossible to plough. If one farmer makes a small drain, then the drainage won't work – it depends on the level of the land. We have to get together, discuss it and get it right (farmer at Tulube, 6 April 1998).

The government and NGOs as sources of knowledge

According to most farmers, the influence of the Ministry of Agriculture (MoA) on wetland farming has been negligible until fairly recently. Although farmers have in the past received information from the MoA on upland farming techniques, little has been directed specifically at wetland utilization until the change in government in 1991. Furthermore, this advice is used selectively by farmers in accordance with their own body of knowledge. As illustrated in Figure 8.8, farmers at Dizi regard the influence of the MoA as significant at the present, where it accounts for approximately two thirds of their total wetland knowledge (represented by sector 1 in the Venn diagram). Farmers suggested that the non-adoption of MoA information occurs because much of this is specific to oxen and not all farmers own, have access to or need draught animals for their farming activities. In addition, Dizi's farmers point out that the MoA tends to overlook the fact that farmers have a problem with wild pests and consequently their advice is often impractical and hence unheeded under the circumstances. Finally, farmers argue that the MoA fails to recognize that they often sell their labour to town merchants in order to raise money to buy food during the hungry season. Investing time and resources in their wetland farming system during this period is often economically unattractive, so again the MoA's advice is not used.

At Supe farmers acknowledge the MoA as a source of new crop varieties and technical advice although they are eager to point out that this advice is often contradictory to that offered by administrative institutions (the *kebele*). If this is the case, then the contradictory information is ignored but if both sources are complementary then it is utilized.

Communication with the MoA was not acknowledged at any of the other study wetlands where PRA was undertaken, although this is not to say that it does not exist where other aspects of the farming system are concerned. Similarly, no farmers mentioned any interaction with NGO staff (MFM are the largest in the area) with regards to wetland utilization and it would appear that their wetland extension activities (and those of the MoA) are limited to a few individual sites, Hurumu having been one of these in the past.

Innovation in the wetlands

Evidence for the communication of wetland knowledge leading to the adoption or modification of practices is perhaps stronger than that which points towards farmers as spontaneous innovators, experimenting with wetland farming techniques. There are however, several examples which hint at this innovative capacity. Farmers at Dizi wetland acknowledged that they were constantly learning from their experiences of wetland cultivation. Experimentation with drainage layout is also an ongoing process among farmers, especially in response to hydrological changes and the problem of inefficient drainage. For example, in one farmer's plot on an eastern limb of Dizi wetland, half of the maize was yellow and half was healthy during the 1997 – 1998 season. The farmer recalled how two years previously there was no difference between them but in the 1996 – 1997 season the farmer lost the whole yield. During the 1997 – 1998 season, the farmers excavated a drain through the centre of his wetland plot, assuming that waterlogging was the problem but this resulted in only half the maize reaching maturity. Advice on drainage was provided by another farmer who told him that the source of water for that area of the wetland was under an avocado tree on the valley side. The farmer claimed that in the following season he would drain along the left side, where the valley sides meet the wetland and he believed that this will solve the problem. In addition, he also intended to block the middle ditch, which was only recently constructed.

This example suggests, however, that although the implications of hydrological changes on drainage regimes are recognized, the rate of adaptation will be slow. Most farmers agree that the depth of their drainage channels should be altered according to the moisture conditions of the soil but in most cases this is carried out only once in a season (excluding general maintenance). Instead of actively managing the depth of drainage channels during the year, changes to the design are carried out in the following season which means that farmers are never able to fully adapt to the hydrological regime, especially during a time in which they describe rainfall as being less predictable.

At other sites such as Bake Chora and Tulube, there are similar indications that water management has developed through a process of trial and error, although no major innovations to their original system were mentioned. In view of the differences in the wetland farming activities it is likely that farmers have developed their methods to suit their individual circumstances.

One minor but obvious adaptation was found at Supe where farmers described modifications which have been made to the *Togo* (machete):

> The handle of the togo has been made longer and stronger wood has been used. For the forest the blade is only sharp on one side but for the wetland both sides are sharp. This is used for ditch clearing and it facilitates the cutting of the fibrous roots of the cheffe. It is also good for when the soil is very wet (farmer at Supe, 22 April 1998).

Another example of this innovation in farming technology is in Bake Chora, where one farmer has constructed human-like scarecrows in wetland plot adjacent to the densely forested valley side which he claims is the main access route for wild pests to the wetland (Figure 8.9). Another is located in his wetland plot where tef is being cultivated. No other farmers within Bake Chora have engaged in this practice to such an extent, although several maintained that they frequently left their clothes around the wetland as a pest deterrent.

One problem which could explain the under-reporting of farmer innovation is the farmers' own perception of what constitutes experimentation and innovation. There is some question over to what extent, when experiments are successful, practices become the norm and are, therefore, not recognized as anything outstanding to farmers. Similarly, farmers may regard small-scale innovations as insignificant and hardly worth mentioning during discussions.

The dynamics of knowledge acquisition

IK is a dynamic and changing phenomenon. As highlighted by many farmers, wetland knowledge has been acquired to different degrees and from different sources since the drainage and cultivation of wetlands was initiated. The only constant source of knowledge which farmers universally recognize is that body of knowledge which they have inherited from their forefathers. This 'original'

Indigenous Wetland Management in Illubabor 177

Figure 8.9   Indigenous pest management technology? A scarecrow in Bake Chora wetland

wetland knowledge is considered by many to be invaluable, along with that held by community elders who are also perceived to have a broader range of experiences of wetland farming.

There is certainly some evidence to suggest that wetland knowledge was acquired much more rapidly and from a variety of sources at the beginning of cultivation than at the present time and that this may have formed the core of wetland knowledge which is applied and modified by farmers today. Several groups of farmers mention that when they were first instructed to cultivate their wetland by the landlords they met and discussed drainage techniques.

During the initial periods of intensive wetland utilization most farmers agree that they did not receive any assistance in terms of technical advice from either the government (MoA) or the private landlords and at this stage the inter-wetland communication of information by farmers was more significant. For example, farmers at Dizi point out that when they initially achieved a good crop yield from their wetland word of this spread to other communities who subsequently adopted their system of wetland cultivation. At Supe the drainage and cultivation technique, accredited to the Koran teacher Sheik Abdulah, was said to be disseminated throughout the area by his students.

With successive changes in government, following the initial periods of wetland cultivation, the mechanisms of acquiring wetland knowledge have also changed. During the Derg regime, wetland farming was disrupted in many areas as a result of increasing demands for farmers to engage in 'State' activities, such as communal work groups. Although directing human resources away from the wetlands these communal groups may have presented an opportunity for wetland knowledge to be transferred between communities. Communication between communities may have also increased as the market economy began to develop further following the downfall of the Derg. In addition, it is likely that, in response to new market forces, farmers will increasingly be required to innovate and alter their wetland farming system towards a state of greater productivity.

Conclusions: farmers' wetland knowledge in context

Farmers' knowledge of wetlands and their utilization is in many ways typical of indigenous knowledge systems found within the developing world in terms of the criteria suggested by various researchers (e.g. Chambers, 1983; Warren, 1991; Rajasekaran, 1993). In terms of wetland hydrological processes, farmers clearly possess extensive knowledge of their natural environment. They have theories regarding the relationship between rainfall and the behaviour of water table and runoff in wetlands, as well as clear ideas of the significance of soil moisture and the influence of this on crop growth. Furthermore, farmers demonstrated very site specific knowledge of factors affecting hydrological functioning and in addition, the spatial and temporal variability of wetland hydrological characteristics. For example, most farmers understood the overall hydrological variation inherent in their wetland and how this affected not only their own wetland plot but also the plots of their fellow farmers.

Farmers' knowledge of their natural environment would, therefore, appear to be extensive, reflecting 'an intimate understanding of the environment in a given culture' (Rajasekaran, 1993). As many of the cited examples suggest, this has been gained through a range of mechanisms the most significant of which have been the farmers' own experiences. There is evidence that farmers have been on a learning curve since they and their ancestors were first involved in wetland utilization. For example, drainage designs appear to have changed in response to crop failure attributed to hydrological problems. These changes, which have included modifications to drainage depth, the position of drains and the tools used to maintain them are a result of observations being acted upon which implies some degree of innovative adaptation. Wetland farming calendars and the specific techniques used are variable between sites, suggesting that some degree of adaptation has taken place since wetland farming began, when traditional upslope cultivation techniques were first applied to the valley bottoms with variable success.

In the broader context there are examples where farmers have learned from their mistakes, for example where *cheffe* is valued above agricultural potential as a result of farmers' experiences of past shortages. Clearly the wetland farmers' knowledge meets the criteria of an IKS in that it is characterized by 'traditional practices which have evolved over time' (Chambers, 1983).

In addition to observation, farmers report that some of their most important knowledge is that which has been passed on by their forefathers and this forms the core of their knowledge system. Their agreement that the influences of external agencies have been negligible in the past, arguably serves to reinforce the evidence that their knowledge and utilization strategies have evolved through 'indigenous mechanisms of creativity and innovativeness' (Warren, 1991). Farmers acknowledge that external actors have only recently had an influence on their wetland utilization strategies and their use of this information has been selective to meet the farmers' individual needs. This represents one area where farmers' wetland knowledge 'reflects experiences based on traditions and more recent experiences with outside technologies' (Haverkort, 1995). It is important to recognize, however, that the participatory nature of this research has also contributed to the farmers' interaction with external knowledge systems, particularly during instances where technical information from the hydrological monitoring programme was shared.

The evidence, therefore, suggests that within the study area, farmers' wetland knowledge shares many characteristics typical of a dynamic, evolving and adapting IKS. Whilst much of the research discussed in Chapter Three provided examples of large scale, conspicuous knowledge acquisition (e.g. Richards, 1985; Chambers *et al.*, 1989; Rhodes and Bebbington, 1995), in this study there were no clear examples of wetland farmers engaging in such formal processes of experimentation. Instead, the small-scale, relatively inconspicuous modifications made to existing practices or equipment appear to represent the main adaptive mechanisms which contribute to the evolution of wetland knowledge and hydrological management.

Farmers at each of the study sites also report a variety of problems which suggests that their current utilization strategies are not completely adapted to their environmental circumstances. Some constraints to hydrological management are physical and the extent to which farmers' wetland knowledge can adapt to these problems is debatable. Furthermore, the number of constraints (both environmental and socio-economic) affecting wetland utilization are such that the wetland farmer arguably has little opportunity to innovate or adapt to his environment. The situation would appear to be typical of that described by Farrington and Martin, (1998) and Ryden (1991) whereby socio-economic changes have exceeded the capacity of farmers to cope and consequently adaptation and innovation is slow.

Placed in context, however, the drainage and cultivation of these wetlands has at most been carried out for 100 years and therefore unlike the majority of documented IK not 'tested over centuries of use' (IIRR, 1996). This may in part, account for the deficiencies in farmers' knowledge acquisition mechanisms and the apparent state of a lack of adaptation to various constraints. It could, however, be argued that the current state of the farmers' wetland knowledge system is all the more remarkable, in view of this relatively short time period.

Whilst agricultural production from these wetlands may ultimately suffer as a result of increasing constraints originating in the current socio-economic climate, the wetland themselves through a lack of adaptive development, may also escape degradation. Removing the current constraints on wetland utilization could potentially have a major impact on the long-term sustainability of the wetland hydrology. Although the farmers' wetland knowledge system potentially has the adaptive capacity (i.e. the mechanisms) to cope with such changes, its success in this instance will depend on the accuracy of their perceptions and understanding of hydrological reality. Hence the next chapter addresses this issue of how farmers' wetland knowledge does compare to the wetland hydrology as described in Chapter Seven and consequently the opportunities for the sustainability of wetland drainage agriculture.

Chapter 9

Indigenous and Scientific Wetland Knowledge

Introduction

IK is now recognized as an important, organized body of knowledge, providing an important basis for natural resource management in the developing world. Evidence collected from various studies around the world suggests that IK, which has evolved over many generations and which has adapted to local cultures and environments, forms the basis of many natural resource management strategies which have also been sustainable for many years. One of the main aims of this research was to identify the characteristics and mechanisms of the indigenous wetland knowledge held by local communities in Illubabor zone, with a view to establishing its contribution to sustainable hydrological management practices.

In the context of the previous discussions on both the hydrological characteristics and the hydrological management of wetlands in the study area, this chapter discusses the relationship between farmers' knowledge of the wetland environment and the hydrological reality. The nature of this relationship can be considered the key to attaining the sustainable hydrological management of these wetlands in that it is farmers' knowledge and the extent to which this reflects hydrological reality, which ultimately determines the ways in which wetlands are used. Consequently, this chapter addresses the extent to which hydrological reality is perceived, understood and applied by wetland users in their management practices. On the basis of this knowledge and its application, the potential of wetland users to achieve sustainable hydrological management is discussed.

Hydrological knowledge and reality

Although the findings of the PRA programme presented in Chapter Eight suggested that farmers possess a significant body of knowledge relating to hydrological processes, it is the extent to which this knowledge represents hydrological reality which ultimately determines the success and sustainability of hydrological management practices. As a means of establishing the accuracy of farmers' understanding of wetland hydrology, their perceptions and ideas as recorded during the PRA sessions were compared to the hydrological data presented in Chapter Seven.

Although seasonal diagramming activities carried out during the PRA programme were designed to facilitate a relative comparison between farmers' perceptions and actual rainfall and water table data, many discussions also focused on a range of other hydrological variables producing qualitative information. Consequently, in addition to the comparison of water table and rainfall graphs, the discussion also explores the relationship between farmers' knowledge and other findings of the hydrological monitoring programme. In particular, the spatial variability of the wetland water table, the hydro-chemical properties of the wetland and the issue of wetland regenerative capacity.

Rainfall

The conversion of meteorological and hydrological records to mean monthly measurements makes them directly comparable with those graphs produced by the farmers at each site, in terms of the number and timing of measurements throughout the year. The actual values from meteorological and hydrological records and farmers' perceptions are not comparable, as farmer perceptions were measured on a purely arbitrary scale. Nonetheless, changes in the pattern of rainfall or water table from month to month are comparable between both data sets.

The seasonal diagrams produced by farmers use the Ethiopian calendar which consists of twelve months of thirty days and one month of five days (CSA, 1997). It was, therefore, necessary to convert farmers observations made in the Ethiopian calendar to the Gregorian (western) calendar in order for farmers' and hydrological data to be comparable. Perceptions of average monthly rainfall were compared to rainfall records from the nearest rain gauge to that wetland.

Farmers' perceptions of rainfall were found to be broadly consistent with rainfall records although each site exhibits its own idiosyncrasies ranging from dramatic fluctuations of rainfall from one month to the next, to the perceived absence of rainfall where meteorological records indicate otherwise. Although the established pattern of rainfall in the region is unimodal, several groups of farmers suggest a bimodal rainfall calendar, exhibiting a second rise and decline of rainfall within the season.

Farmers at Bake Chora and both groups at Tulube suggest a rise in rainfall during November, although these perceptions tend to conflict with meteorological records (Figures 9.1b, 9.1e and 9.1f). In these cases, however, it is possible that in the construction of the seasonal diagrams, farmers demonstrated knowledge of the immediately preceding year's rainfall rather than highlighting average conditions. During the 1997 – 1998 season, October received more rainfall than both September and November and, when considered in the context of the Ethiopian calendar, the farmers' peak during November mirrors the reality of this abnormal rainfall event.

Another inconsistency between farmers' perceptions and the meteorological data is where farmers indicate a complete absence of rainfall where records suggest the contrary. This is the case for both Dizi (Figure 9.1c) and Tulube (group 2) (Figure 9.1f) during February. Gill (1991) suggests that such inconsistencies can be explained by the fact that farmers view rainfall within their own subjective frame

Indigenous and Scientific Wetland Knowledge

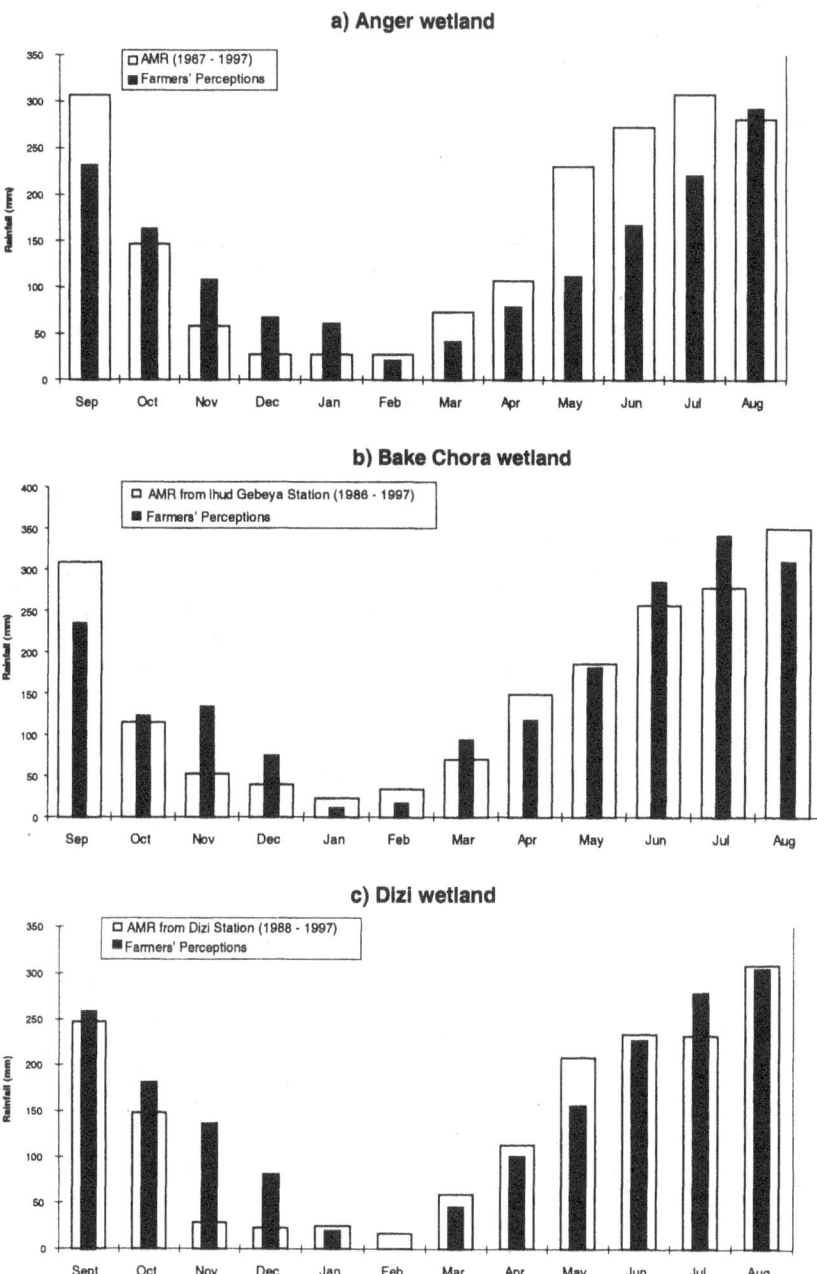

Figure 9.1 Farmers' perceptions of rainfall compared to rainfall records

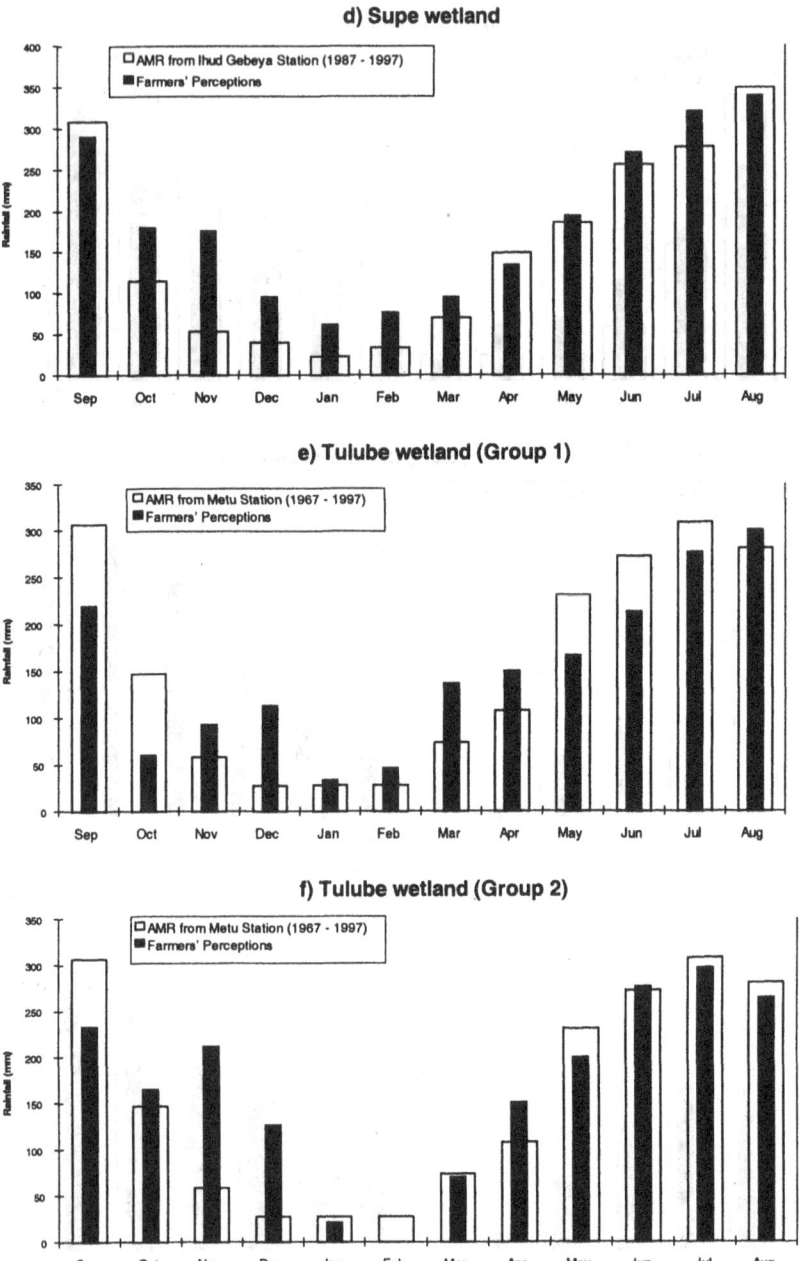

Figure 9.1 (cont.) Farmers' perceptions of rainfall compared to rainfall records

of reference. In other words, farmers see rainfall as inherently linked to agricultural productivity, to the extent that its significance (and therefore its perception) changes depending on the time of year or the farming activity in hand. In this respect, the failure of farmers to notice any rainfall (which may amount to less than 20 mm anyway) at the height of the dry season and during a period in the farming calendar when no farm activities are undertaken, is understandable. Later in the season, rainfall becomes more significant as crop yield becomes more sensitive to moisture conditions and during this time farmers' perception of changes in rainfall may be heightened.

A potential problem with the interpretation of these data is determining what rainfall characteristics the farmers actually represented on the graphs they produced. Although the graphs were intended to represent a measure of the volume of rainfall which fell during each month, studies have shown that this is often confused (by both farmers and researchers) with the number of days in which rainfall fell during a specific month (Gill, 1991; Chambers, 1994b). Furthermore, it should not be assumed that farmers conceptualize rainfall in terms of either volume or duration as is the scientific norm. The inconsistencies between both data sets may simply reflect the differences between the knowledge systems from which each originated, whereby the farmers' knowledge is rooted in their experiences in a particular environment whilst the meteorological data represents an absolute value for a specific location over a finite time period. The significance of rainfall to farmers lies in its effect on crop production and, therefore, farmers are more likely to perceive and recall rainfall which has implications for their way of life. Bearing in mind these differences of interpretation and the inaccuracies produced through the conversion of data between calendars, the similarity between perceptions and the meteorological reality remains striking.

Seasonal changes in water table levels

A comparison of the monthly mean water table levels and farmers' perceptions of seasonal changes in water table levels is presented in Figure 9.2. This clearly shows an overall dissimilarity between water table records and how farmers perceive the wetland water table relative to the surface of their wetlands during the year. The most obvious source of dissimilarity between the hydrological data and farmers' perceptions is that farmers perceive the water table elevation as much higher than the hydrological data suggests. This is clearly the case in Anger (Figure 9.2a) and Dizi (Figure 9.2c) wetlands, and to a lesser extent in Tulube (Figures 9.2d to 9.2f) and Bake Chora (Figure 9.2b). In Anger wetland this can be explained by the fact that those farmers attending the PRA sessions were not cultivating their wetland plots and consequently their view of water table elevation reflects that of natural conditions (whereas the real data represents an area of cultivated and degraded wetland). In Tulube (Figures 9.2e and 9.2f), however, the whole of the wetland is under regenerating conditions and, whilst the hydrological data show a water table which is predominantly below the surface for much of the year, farmers suggest that water remains above the surface throughout the year. Similarly at Dizi

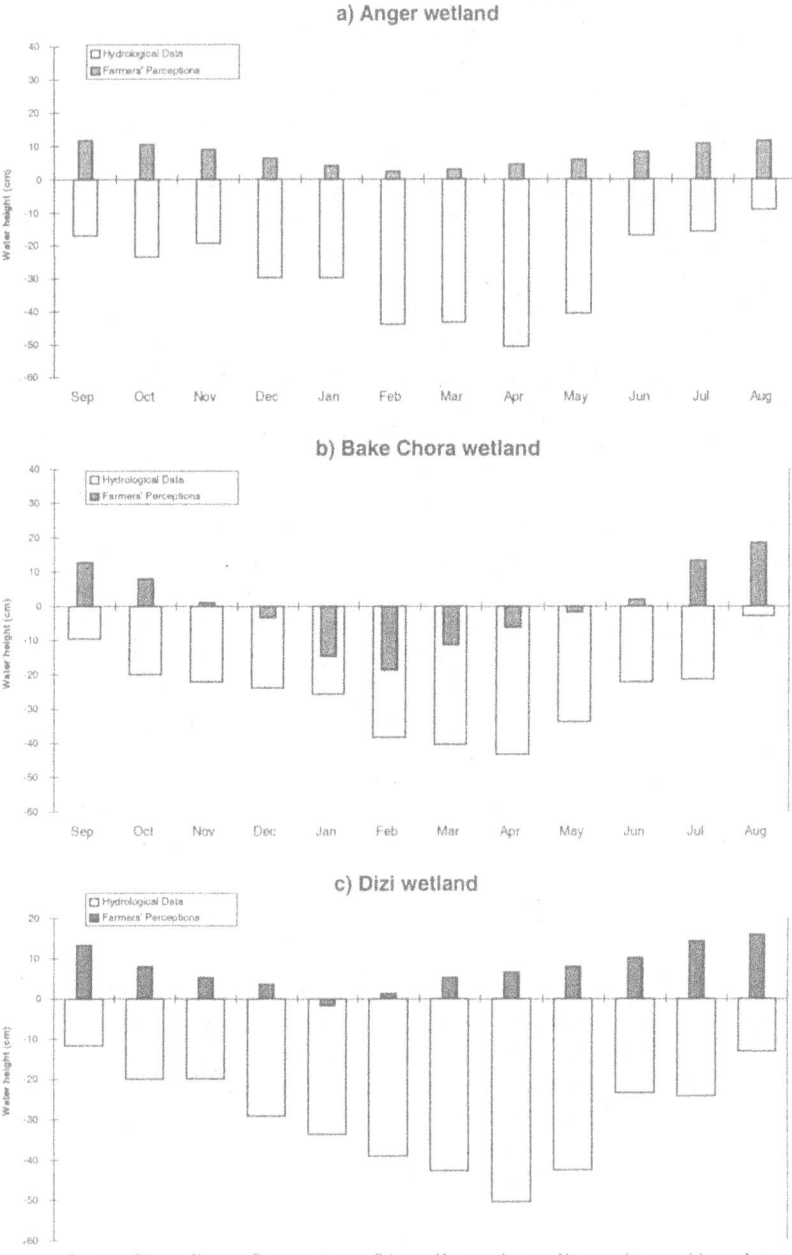

Figure 9.2   Farmers' perceptions of water table elevation compared to hydrological records

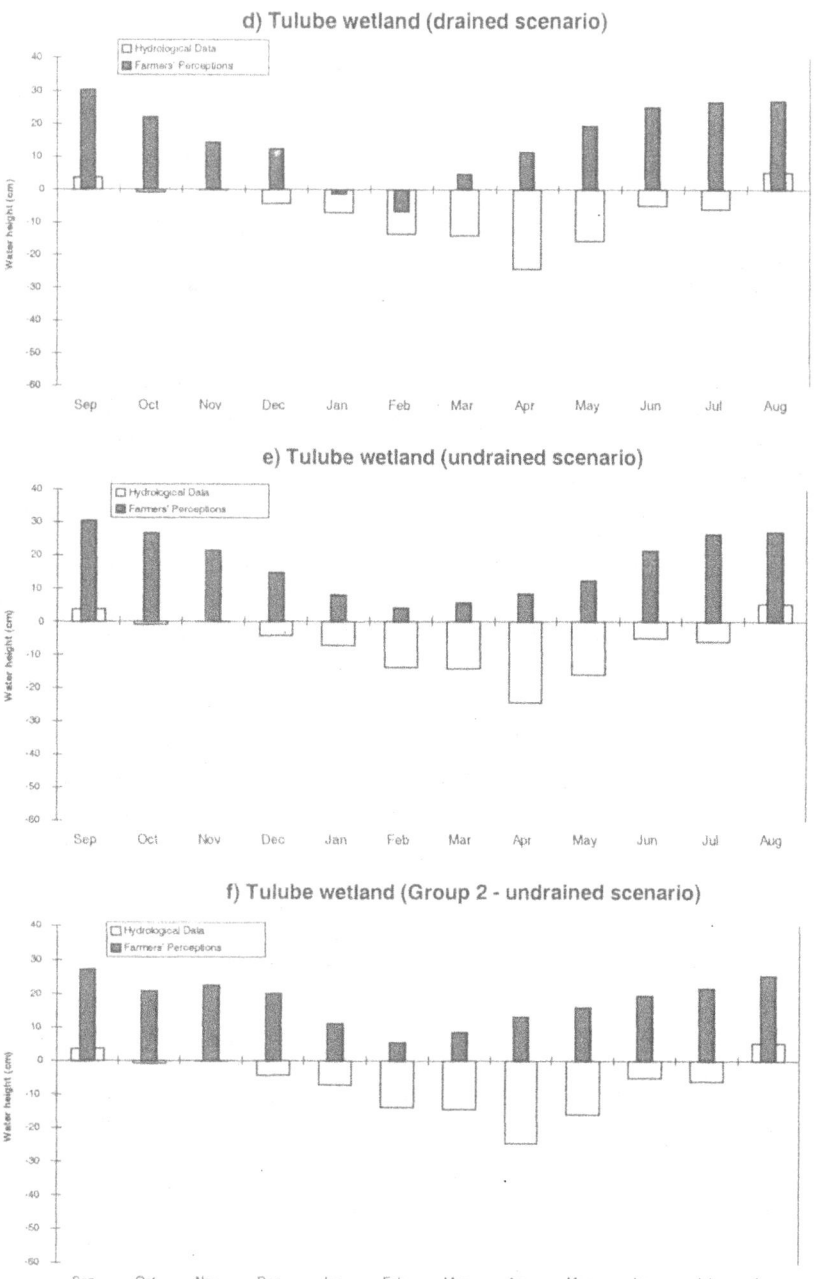

Figure 9.2 (cont.) Farmers' perceptions of water table elevation compared to hydrological records

(Figure 9.2c), which remains fully cultivated throughout the year, farmers indicated that the water table level remains above the surface of the wetland. Only at Bake Chora (Figure 9.2b) do farmers' perceptions resemble that of the hydrological data, although perceptions contradict the hydrological data for six months of the year.

Despite the apparent distorted perceptions of water table height throughout the year, farmers demonstrated a high degree of accuracy in their knowledge of the relative changes in water table height between each month. Figure 9.3 shows an overlay of the farmers' perceptions of water table height with the hydrological data for each site, illustrating the changes occurring during one year.

The general pattern of water table behaviour described by farmers is one of a decline in water table elevation to a minimum level during January and February. The hydrological records, however, show a minimum water table height occurring much later in the season between April and May. This discrepancy between the two sources of data can be explained by the anomalous weather conditions affecting south-west Ethiopia during the 1997 – 1998 season. These produced an extension of the dry season, delaying the onset of the rains and ultimately resulted in an abnormal lowering of the water table throughout the study area. The implications of this are evident particularly in Anger wetland (Figure 9.3a) where, in reality, there is a marked fall in the water table during April, whilst farmers' knowledge would suggest a steady rise from March. Similarly at Dizi (Figure 9.3c) the influence of the late rains on the water table is clear, as is the higher rainfall experienced during November 1997 and the lower rainfall during July 1998. These contrast with the farmers' perceptions of water table variation which represent 'normal' conditions and arguably a lifetime's experience.

In conclusion, taking into consideration the anomalous weather conditions during the study period, the data presented (Figure 9.3) suggests that farmers are able to predict with great accuracy the relative changes in water table elevation in their particular wetland during the year.

With respect to the dissimilarity which does exist between the recorded water table levels and farmers' perceptions, an important consideration is whether the knowledge of those farmers who took part in the PRA sessions was representative of the instrumented areas of each wetland (which constituted the headland area of most wetlands). For example, those wetland plots owned by the participants at Anger were not covered by the hydrological monitoring programme, rendering any comparison between the two misleading (although interesting in that the farmers' perceptions confirm the expected pattern of water table in a *cheffe* dominated area of that wetland). Similarly in Bake Chora, the hydrological data are representative of only the headwater area of the wetland which is exclusively owned by one farmer, although the diagrams constructed were the result of a general consensus among several farmers on water table levels throughout the wetland.

The seasonal diagram activities were carried out to establish farmers' knowledge of normal hydrological conditions in the wetland. The data generated, therefore, represents an average of their experiences whereas the hydrological data is limited to one season. This does not, however, necessarily imply that farmers are inexperienced with hydrological deviations. Many participants were able to

Indigenous and Scientific Wetland Knowledge 189

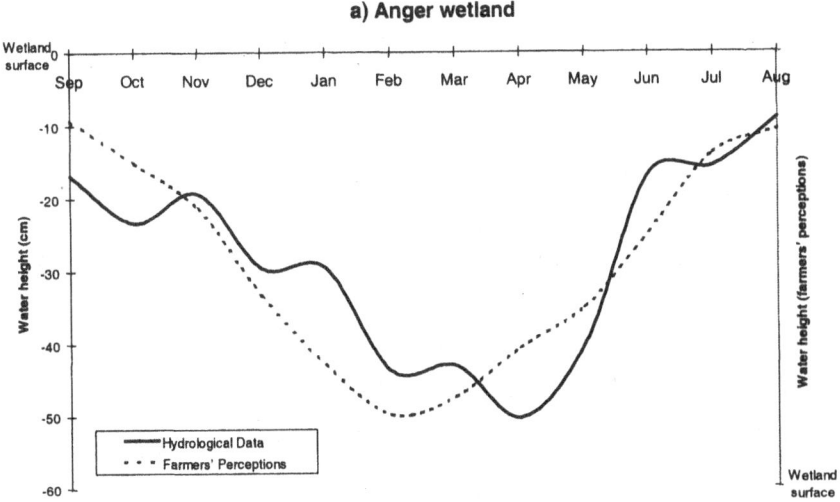

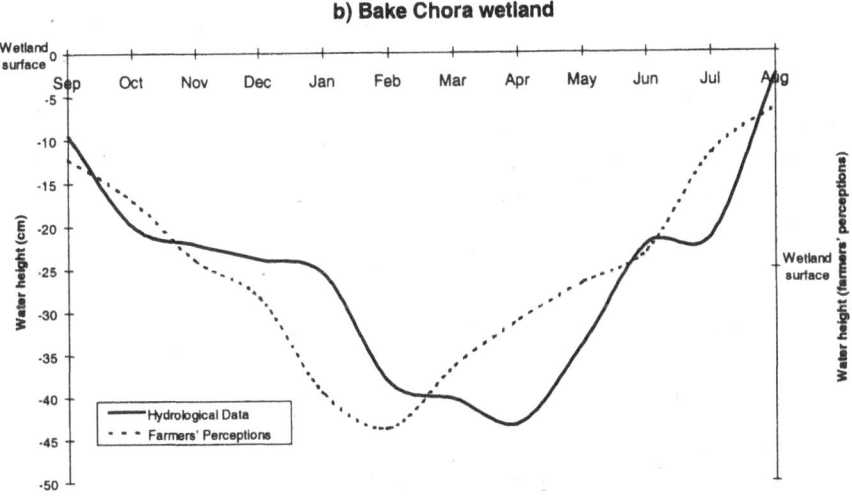

Figure 9.3 Farmers' perceptions of water table compared to the hydrological data

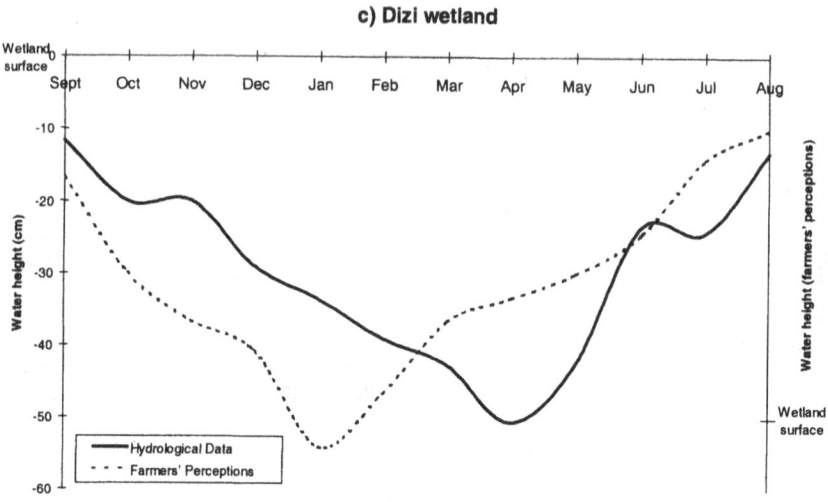

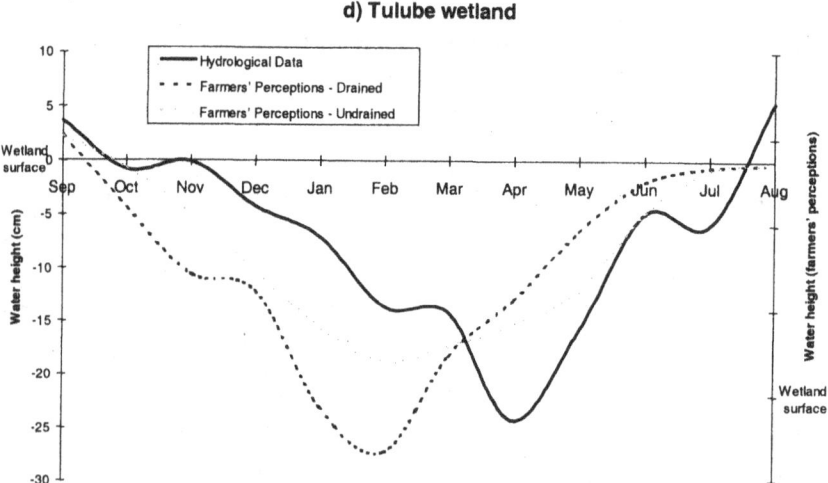

Figure 9.3 (cont.) Farmers' perceptions of water table compared to the hydrological data

describe in detail the effects of, for example, delayed rains on their wetland crop.

Whilst the spatial and temporal variations between farmers' perceptions and the hydrological data may explain the dissimilarity, there are potentially cultural differences in concepts such as 'water table' and 'surface'. Given that in most cases farmers show the water table to be above the surface of the wetland when evidence suggests it is still below the ground, it is logical to assume that the problem is one of calibrating the farmers' view of the wetland surface with the hydrological reality. As both the perceptions and the hydrological data follow similar trends during the year this suggests that farmers do have an understanding of the changes in levels taking place, albeit rather transposed. It is possible, therefore, that farmers use a reference point different to the wetland surface as identified by this research, from which changes in water table height are observed. On the basis of the comparisons made in Figure 9.3 it is likely that one such reference point could include the bottom of drainage channels.

Finally, it is necessary to understand that hydrological management is an emotive subject for farmers. Drainage and the maintenance of the system throughout the farming calendar is extremely labour intensive whilst the returns on labour are often low. Within this scenario, water table height is inevitably of primary concern to farmers who may be reluctant to underestimate its influence, even when their drainage efforts are reasonably efficient.

Spatial variability in water table dynamics

The wetland typology discussed in Chapter Seven highlighted the existence of some degree of variability in the behaviour and characteristics of the wetland water table within each study site. The typology also demonstrated that some wetlands possessed more spatial variability in terms of their water table behaviour than others. In particular, Chebere and Tulube wetlands were found to have more homogeneous water tables in comparison to other sites, especially Anger and Hurumu where the behaviour of the water table during the 1997 – 1998 season was found to be variable within the short distances between dipwells. This variability was, to some extent, also confirmed by the results of the hydraulic conductivity investigations which suggested that the wetland soil's ability to transmit water was more spatially variable in those wetlands which had recently, or were at that present time, undergoing drainage and cultivation. Such wetlands were also more likely to demonstrate a much lower range of hydraulic conductivities, which are potentially the result of physical and chemical changes taking place in the soil following several years of drainage and cultivation.

It has been suggested that this spatial variability in water table characteristics has direct consequences on the land use within a wetland. Given this variability, it is unlikely that management practices which are designed to suit a particular hydrological regime can be effective throughout each wetland as a whole. This appears to be the case in wetlands such as Bake Chora, Dizi and Anger where the hydrological data suggests that some areas are well drained throughout the growing season (the water table is maintained below the root zone) whilst others suffer from waterlogging which induces crop damage. Where the latter occurs, farmers may be forced to abandon their wetland plots where returns to labour are low. This can lead to a spatially variable land use within the wetland which may, in the long-term, exacerbate the spatial variability of water table characteristics.

Many farmers during the PRA sessions demonstrated knowledge of the spatial variability of the wetland water table which complements and adds to the findings of the hydrological data. Most farmers agreed that the wetland water table varied in height throughout their wetlands and some highlighted particular areas where waterlogging prevented crop cultivation. In addition, farmers also suggested that the influence of several factors affected the characteristics of the wetland water table at a particular location:

1. the presence of rocks near the surface of the wetland was regarded as an obstruction to the flow of water, thereby raising the water table;
2. sandy soil was offered as an explanation for a low water table in one area of Dizi wetland;
3. the slope of the wetland was said to affect the flow of water, with steeper 'uneven' areas having lower water tables than flat areas;
4. inadequate drainage was a also regarded as a cause of water table variation. Farmers at Bake Chora pointed out that abandoned areas of wetland where drains were not maintained would affect the water table of the surrounding plots.

Relating these to actual hydrological measurements is problematic in that in most cases farmers identified areas of water table variation which were located outside of the dipwell transect. Nonetheless, it is clear that the variation highlighted by the water table typology does, on the basis of farmers' knowledge, exist throughout most wetlands. Some direct comparison between the two sources of information can, however, be made with respect to Bake Chora wetland where a transect walk was undertaken. Farmers pointed out that an area of discoloured maize near dipwells 9 and 10 was a result of the presence of excess water in the soil. According to the water table typology, both dipwells represent areas where the water table was found to be particularly high in comparison to most of the others at this site (Figure 9.4), confirming the farmers' observations.

A similar situation exists at Dizi in that farmers drew attention to wetter and drier areas within the wetland. Although these areas were not covered by the dipwell transect, the variability in the water table typology for this site does indicate a range of different drainage conditions within the wetland. Again, farmers would appear to be knowledgeable of this variation and its implications for wetland use.

Regeneration

The issue of wetland regeneration was one of the key themes to emerge from both the results of the hydrological monitoring programme and the PRA discussions. In terms of their water table hydrology, the regenerating site of Tulube was found to share similar hydrological characteristics with both Chebere and upper Wangeneye

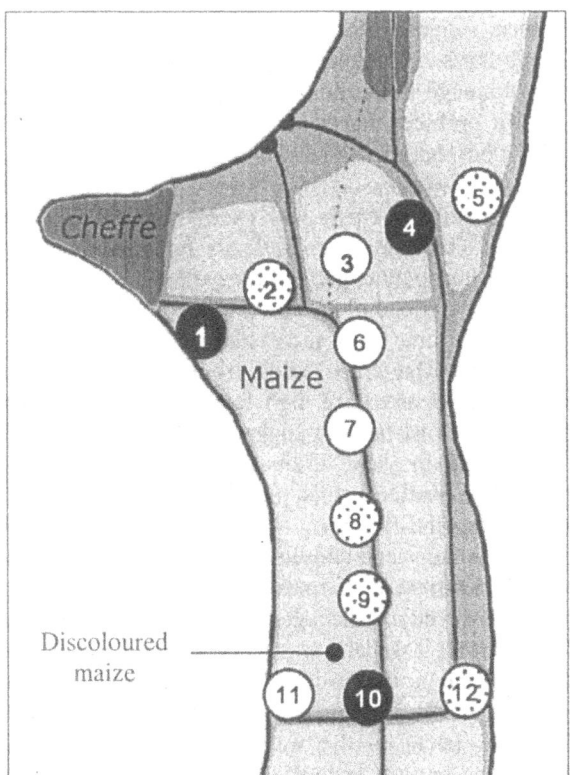

Figure 9.4   The water table typology in Bake Chora wetland and the location of an area of discoloured maize

in that a high, less erratic water table was maintained for much of the year. The hydraulic conductivity measurements also suggested that the movement of water through these wetland soils was very rapid and uniform in contrast to other wetlands where K_{sat} tended to be slower and more spatially variable. This situation was attributed to the recolonization of wetland vegetation in these abandoned areas, which facilitated sedimentation, the build up of organic matter and the improvement of soil structure.

Although the hydrological data suggests that these wetlands may be more robust than previously expected, there is clearly a need for further hydrological investigations into the regeneration process. Farmers' knowledge of wetland regeneration, however, was found to be part of a wider detailed knowledge of ecohydrological interactions within the wetland environment. This knowledge verified the findings of the hydrological monitoring programme but in addition, farmers were also able to demonstrate a much greater understanding of the wetland regeneration process and the subsequent recovery of natural hydrological conditions.

Critically, farmers suggested that following the abandonment of a wetland or wetland plots, *cheffe* starts to recolonize rapidly. Even when the wetland is drained, those areas where drainage is inefficient are characterized by *cheffe* growth and this is regarded as a problem. Farmers suggested that excess water creates the conditions for *cheffe* but the growth of *cheffe* also facilitates a rise in the wetland water table. Although *cheffe* and its waterlogged conditions can be problematic, knowledge of the association of *cheffe* with excess water is also used by farmers to their advantage. In particular, areas of *cheffe* are regarded as reservoirs which can maintain a supply of water to other parts of the wetland.

The abandonment of wetland plots, characterized by the regeneration of *cheffe*, is also used by farmers as a means of increasing soil fertility in that particular area. Farmers also report that the crop yields from newly drained and cultivated wetlands (after *cheffe* is cleared) are high for the first few seasons, after which yields start to decline. Whilst farmers suggested that pest problems may account for some of this decline in yield, drainage and cultivation inevitably lead to physical changes in the wetland soils, notably compaction and mineralization, which subsequently affect the efficiency of drainage.

Overall, the characteristics and consequences of abandonment and regeneration appear to be well understood by farmers, to the extent that they constitute an important component of wetland management strategies. Although the short time period in which hydrological data was collected was insufficient to provide supporting evidence of the processes described by farmers, the data collected from the various wetlands (which do represent different development stages) suggests that regeneration does occur in the ways farmers described. This has major implications for the long-term hydrological sustainability of these wetlands, based on indigenous knowledge.

Hydrochemistry

Farmers were found to possess knowledge of the seasonal changes in water colour within their wetland and also how this is linked to rainfall and catchment runoff. Descriptions of the variations in water colour produced by several groups of farmers suggest that there is some variation in the wetland hydro-chemistry during the year and also between different parts of a wetland. The commonly held view was that these variations were linked to rainfall and soil erosion in the catchment. Given these variations, it would be expected that the perceived changes in water colour are indicative of chemical changes in the wetland runoff originating either from the leaching of wetland soils or from the transport of sediment from the catchment. The results of the hydrological monitoring programme, however, remain inconclusive with respect to seasonal and spatial changes in nutrient levels and no clear pattern was observed.

On the basis of their knowledge on other hydrological variables, it is likely that the changes in water colour described by farmers do represent real changes in the hydro-chemical characteristics of the wetlands. The hydrological data are not representative of these conditions potentially as a result of limitations in the hydro-chemical monitoring programme, in particular the relatively long time interval between sampling, the location of the water sampling points and the constant problems encountered with monitoring equipment. Farmers' perceptions of water colour changes are in addition, very specific, which lends credibility to the validity of their observations and overall, their knowledge of these processes.

Farmers' knowledge and reality reviewed

This comparison between the hydrological data and farmers' wetland knowledge has highlighted a relationship between scientific and indigenous knowledge which is not dissimilar to that reported by Richards (1980). Whilst both sources of knowledge complement each on some aspects of wetland hydrology, in other cases the knowledge held by farmers goes beyond that generated by scientific monitoring and vice versa (Figure 9.5). The value of the scientific knowledge is evident with respect to the precise measurements of water table elevation during the study period. Although farmers' perceptions of seasonal water table changes did in general follow the trend highlighted by the hydrological data, there were some differences in terms of the general level at which water table variations were occurring. In this case scientific knowledge transcends IK in that its data accurately represents the hydrological conditions in a particular wetland at a particular time. In addition, although farmers did draw attention to the variability of the water table within their particular wetlands, the hydraulic conductivity data presents a means of establishing in more detail, the nature of this variability.

Farmers' knowledge and the hydrological data tended to complement each other in areas such as seasonal rainfall changes and the spatial variability of the wetland water table. In the latter, the typology produced from actual water table monitoring suggested a variability in water table characteristics which was described on many occasions by the farmers themselves in terms of waterlogged or dry wetland areas. The farmers were, however, able to demonstrate a much greater understanding of the relationships between water table variation and the regeneration of natural wetland characteristics, in a way which the hydrological monitoring programme could not establish. Furthermore, the farmers' knowledge of the changes in water colour is also more extensive than any hydro-chemical relationship suggested by the hydrological data. Whilst a reliance on hydrological data alone could have produced misleading and unrepresentative results, setting this within the context of farmers' hydrological knowledge has, at the very least, drawn attention to the hydrological and meteorological anomalies of the study period and the normal conditions under which wetland utilization takes place. The comparison between the two sources of data which are representative of two established knowledge systems, also highlights the limitations of relying on one particular data source. In the context of the wider debate on rural development, these investigations clearly demonstrate that a greater understanding of natural resource management issues can be attained if research and development adopts an approach which incorporates a variety of knowledge systems.

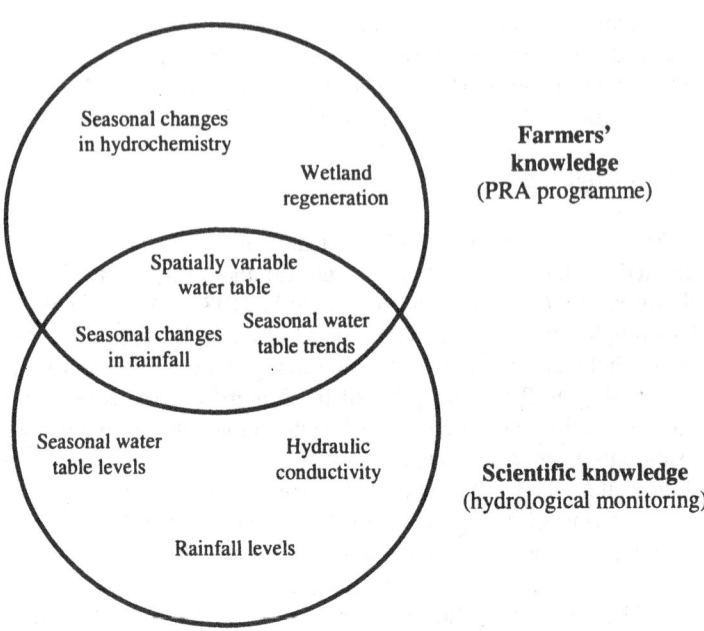

Figure 9.5 The relationship between farmers' wetland knowledge and that generated by hydrological monitoring

Translating knowledge into practice

On the basis of the research findings presented, farmers would appear to possess a knowledge of their wetland hydrological environment which is both extensive and, when compared to the scientific hydrological data, broadly accurate. This knowledge offers a rich degree of contextuality which is based on farmers' experiences of hydrological management over many seasons in their particular wetlands. In view of this experience and the knowledge which farmers demonstrated, it could be expected that the application of this knowledge would provide a basis for successful and sustainable hydrological management, whereby farmers are able to fulfil their goals whilst ensuring that their hydrological management activities do not degrade the wetland. The extent to which this knowledge is used and applied and whether the resulting management interventions can form the basis of the sustainable hydrological management of these wetlands, is now addressed.

The application of hydrological knowledge

The key hydrological management activities in which hydrological knowledge is utilized in the study wetlands include the excavation, maintenance and blocking of drains in an attempt to manage the water table level so that it facilitates the cultivation of crops. In addition, there are some instances where farmers manage *cheffe* vegetation specifically for its water retention capacity. Figure 9.6 illustrates the timing of these hydrological management activities at Bake Chora, a fully cultivated wetland, in relation to the mean weekly water table elevation there and that of Chebere, a pristine wetland, which is included for comparative purposes.

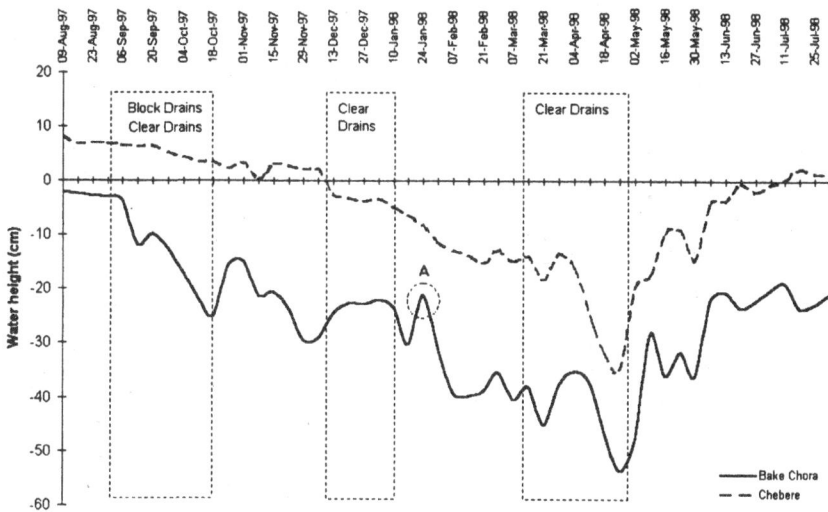

Figure 9.6 **The mean weekly water table in Bake Chora wetland and the timing of farmers' main hydrological management activities**

Clearly in its drained state, Bake Chora's water table has been lowered although in general it shows a similar seasonal trend to that of Chebere which is a result of the influence of rainfall. Whilst there are some instances where the fluctuations in Bake Chora's water table differs from that of Chebere, these do not necessarily correspond to the specified human interventions. Furthermore, during the periods in which active hydrological management is taking place, the effect on the water table appears to be inconsistent. For example, the clearance of drains during December corresponds to an increase in water table height and blocking and clearance activities between September and October are both characterized by a sharp decrease in water height. Of the remaining periods, fluctuations can be attributed to rainfall on the grounds that Chebere exhibits a similar (although not identical) behavioural response.

Although the rise at point 'A' suggests a site specific influence, subsequent analysis of the individual dipwell time series revealed that this rise is a result of a distortion of the mean value influenced by a high water table elevation in one dipwell and missing data for several dipwells located in the drier part of the wetland. In comparing the behaviour of individual dipwells to the hydrological management activities, its effects are equally ambiguous although with respect to activity during September and October the trend in most dipwells is typical of that expected where an area of land is drained causing a fall in water levels.

A sharp increase in water levels towards the end of this period could also be expected following the blockage of drains which would increase water levels behind the blockage. Similarly as the drains are unblocked and cleared during December the data suggests a temporary recovery in water levels although this is far from conclusive. Drain clearance during April would appear to have a direct effect although this is difficult to establish in view of the drought conditions during this period. Although the hydrological management activities do seem to have some effect on water table levels, it appears to be extremely variable in its influence.

A similar comparison can be made with Dizi wetland. The farmers specified ditch clearing during September as the only major structural alteration to drains undertaken within the wetland farming calendar. During the period in which this activity was undertaken, the water table level shows an initial increase followed by a gradual decline (Figure 9.7) although this does not stand out as significant compared with the observed water table fluctuations during the remainder of the year. In general, however, the effects of drainage are still evident in terms of the low and variable water table. It is surprising, therefore, that farmers at Dizi report numerous problems associated with drainage and that throughout the wetland the maize crop is of varying quality depending upon the soil moisture conditions at particular locations.

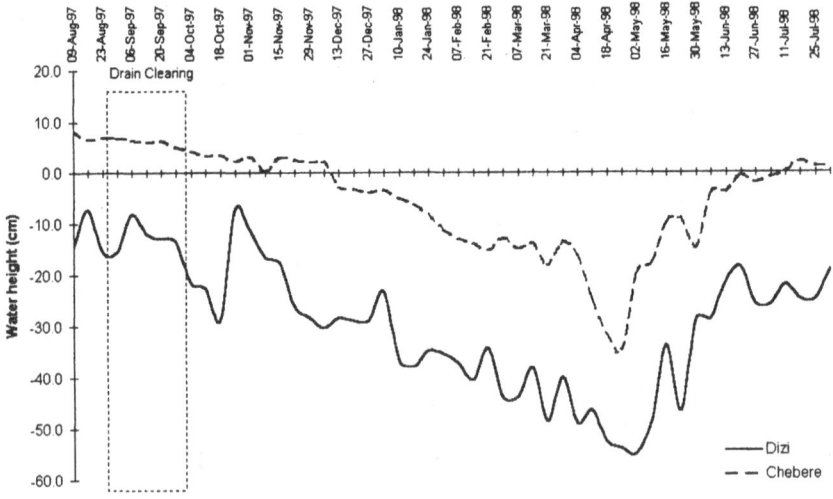

Figure 9.7 The mean weekly water table at Dizi wetland and the timing of farmers' main hydrological management activities

Although PRA sessions were not undertaken at Wangeneye wetland, records of wetland activities were kept during site visits during the study period. These, together with data obtained from Tegegne Sishaw (1998) provide some indication of hydrological management in the wetland during the year (Figure 9.8). Ditch clearing is reported to occur during August and ditch blocking is undertaken when necessary in order to increase the soil moisture content following sowing in January (Tegegne Sishaw, 1998). For the 1997 – 1998 season ditch excavation was recorded at the end of April and July 1998. When plotted in relation to water table levels, each of these periods coincide with a rise in water height which is arguably the opposite of the expected behaviour (Figure 9.9). A possible explanation, however, is that the excavation and clearing of drains results in a flush of water from the large area of *cheffe* upstream. This may cause the temporary increase in water table observed in the data. In general, like Bake Chora and Dizi, water table levels remain relatively low in Wangeneye throughout the year, although the effects of hydrological management are not uniform throughout the wetland.

This confirms that the effects of hydrological management are spatially variable and consequently present problems for crop production. The data from Bake Chora suggests that despite the farmers' efforts at hydrological management, the areas represented by dipwells 1 and 4 in particular are difficult to drain. These areas have consistently high water tables especially during the last four months of the maize growing season, which would prove disastrous for the crop if the root zone were saturated for more than 24 hours (Acland, 1971). In contrast, the drainage regime at both Dizi and Wangeneye would appear to be more successful in that the areas represented by the dipwell transects show a lower water table throughout the growing season. In terms of crop production, however, farmers at Dizi agree that some areas within this wetland facilitate maize cultivation better than others and that these are determined by the water table.

Despite the range of hydrological problems faced by farmers, which are largely a result of over or under-drainage, there is little evidence to suggest that it is a lack of knowledge which is restricting farmers from overcoming these constraints. The design and excavation of drainage networks is not carried out in a haphazard fashion. Farmers appear to be knowledgeable on a range of hydrological processes, acknowledging that their drainage systems are designed according to their previous experiences of the wetland environment and modified each year depending upon recent climatic trends and their experience of the year before. Given this level of spatial awareness combined with the accuracy of their knowledge of hydrological trends, the expected outcome would be an efficient and successful system of hydrological management which realizes their goals in terms of sustaining the conditions which facilitate annual crop production.

In reality, however, this is not always the case and crops are often damaged because the effects of hydrological management are not evenly distributed throughout each wetland. This suggests that for various reasons, the application of farmers' wetland knowledge is not complete.

200 Indigenous Management of Wetlands

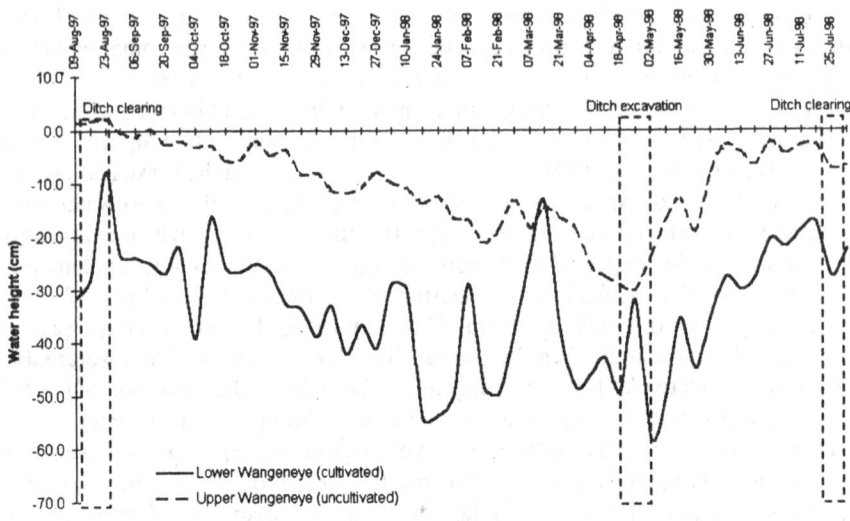

Figure 9.8 The mean weekly water table at Wangeneye wetland and the timing of farmers' main hydrological management activities

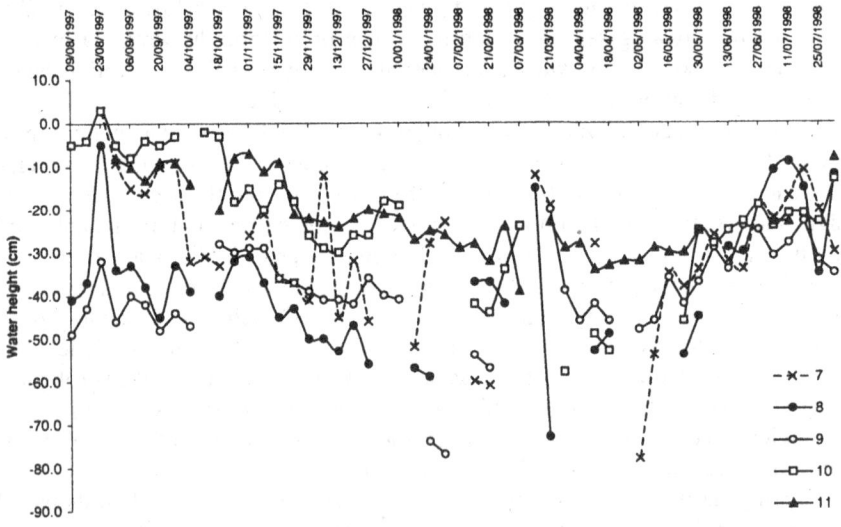

Figure 9.9 Weekly water table levels in lower Wangeneye wetland

The non-application of hydrological knowledge

On the basis of the discussion on hydrological degradation it could be argued that the lack of hydrological degradation is an indicator that hydrological management is, in each case, well adapted to site specific conditions. It has, however, also been pointed out that despite a lack of hydrological degradation and the vast body of hydrological knowledge possessed by farmers, the situation within the study wetlands is one where hydrological management practices do not always and uniformly fulfil the goals of the farmers, and poor crop yields are attributed by farmers in part to inadequate drainage conditions.

One of the key themes to emerge from the data collected during the PRA sessions, is that the whole system of wetland farming is extremely difficult for local communities. Those engaged in the hydrological management of wetlands are operating under a range of constraining influences both at the wider socio-economic and political level, and at the level of their individual farming plots. The major constraints which appear to affect the application and success of hydrological management include a degree of climatic uncertainty, existing geomorphological constraints in the wetlands and the socio-economic situation of the farmers.

Climatic uncertainty The most universal limiting factor is that of the unpredictability of hydro-meteorological conditions. On several occasions farmers stated that in recent years the weather had become unpredictable and the weather during the 1997 – 1998 study period was not representative of normal conditions.

The impact of this unpredictability has been that many of the activities developed over the years within the wetland farming calendar are no longer ideally adapted to hydrological conditions – one farmer in particular at Bake Chora expressed concern over the delay of the rains during 1998, fearing that drought would precipitate an outbreak of disease in his maize crop. More importantly, hydrological management practices have also been designed to cope with expected levels of rainfall and runoff during specific periods. Although farmers do have experience of the effects of hydrological variations throughout the year, their knowledge is of limited use in forecasting climatic anomalies which may occur throughout the wetland farming season. Moreover, the extent to which the effects of abnormal hydrological conditions can be mitigated is limited, particularly when drainage design and maintenance is carried out so early in the season.

Whilst it could be argued that farmers have the capacity to make alterations to their hydrological management throughout the year it should be stressed that they also have a range of other equally important tasks requiring their attention both in the wetland and on their interfluves, e.g. farming, labouring, coffee production, market days. Only at Wangeneye and Bake Chora was there any evidence of farmers undertaking the management of soil moisture conditions (i.e. ditch blocking) later in the farming season in response to rainfall conditions. At other sites, resources appear to be diverted away from hydrological management activities once the wetland has been prepared for cultivation. Hydrological management of wetlands is only a small part of a wider food production strategy and hydrological unpredictability potentially pushes the farmers' capacity to cope to the limit.

Geomorphological constraints In contrast to climatic uncertainty, there are physical, geomorphological features within the wetlands which consistently influence wetland hydrology. These features are regarded by farmers as permanent problems which cannot be overcome by any form of hydrological management. The most commonly cited problem was that of rocky outcrops which either limit the drainage depth or reduce the area of cultivable land. Other factors such as the slope of the wetland surface and the valley sides also play a role in determining the drainage characteristics within each wetland.

These influences together with the effects of climatic variability potentially produce a wide range of possible hydrological scenarios both spatially and temporally, rendering complete adaptation virtually impossible within the context of current drainage and cultivation practices.

The farmers' socio-economic situation The most influential factor affecting the management of wetlands at the present time would appear to be the farmers' socio-economic situation. The wealth and resources of each farmer determine the extent to which they utilize the wetlands for food production or natural products, but also the way in which they use the wetland (Afework Hailu, 1998; Tegegne Sishaw, 1998; Solomon Mulugeta, 1999). Wealthy farmers can afford the benefits of cattle, labour and farming equipment which significantly affect the level of agricultural output. They may also have little need for *cheffe* as a roofing material if wealthy enough to afford corrugated iron.

In contrast, poorer farmers may have insufficient resources to effectively cultivate their wetland plots and, as demonstrated at Bake Chora, may provide the labour for other wetland farmers. Low socio-economic status is a constraint both directly in terms of the wealth and resources available for effective wetland cultivation, and indirectly in that one farmer's use of his wetland plot inevitably affects the adjacent farmers. For example, the abandonment of wetland plots which revert to *cheffe* can have serious effects on the hydrology of adjacent cultivated areas and in addition, the extent to which the wetland farming community are able to co-ordinate their hydrological management activities.

In addition to these constraints which directly affect the ways in which farmers are able to apply their wetland knowledge, it is important to acknowledge that in some cases lack of knowledge may be a constraint to effective and sustainable hydrological management. Although the farmers attending the PRA sessions did demonstrate knowledge of a variety of hydrological processes and wetland management issues, it is possible that not all wetland community members possess the same knowledge. One of the limitations of PRA is that by definition it is based upon the participation of all the group members and the outcome is usually based upon the consensus of each group. It should be recognized that this research represents an initial exploration into farmers' wetland knowledge and there is a need to triangulate the information collected and identify how hydrological knowledge is distributed among the participants.

As suggested by several authors (Swift, 1979; Mundy and Compton, 1995), different members of a community may be skilled in particular ways and possess knowledge of different techniques. Even where a particular practice and the theory on which it is based is common knowledge, there may be differing opinions over how or when it should be applied. At Bake Chora for example, several farmers pointed out that a fellow farmer had not maintained his drain correctly and he had cleared it too late into the wet season so that his crop was damaged. Whilst the constraints discussed above may have accounted for the Bake Chora farmer's behaviour, it is also possible that his drainage was carried out to the best of his knowledge. Another example is provided by the seasonal wetland calendars, where there is some variation between wetlands in terms of the number and timing of wetland management activities.

In summary, the constraints outlined above appear to have a major influence on the ways in which farmers' hydrological knowledge is applied as hydrological management practices. In some cases they may totally hinder the use of hydrological management practices whilst in others the effect may be a reduction in the effectiveness of hydrological management practices or strategies, as is evident throughout the study wetlands.

Conclusions

It has been confirmed that farmers possess an extensive body of knowledge relating to wetland hydrological processes. By highlighting the relationship between this knowledge and that generated by a scientific hydrological monitoring programme, this chapter has attempted to assess the validity and accuracy of farmers' hydrological knowledge. Farmers clearly possess knowledge of seasonal water table trends, seasonal changes in rainfall and the spatial variability of the wetland water table, which reflect the findings of the hydrological data. Farmers also demonstrated knowledge of the dynamics of wetland regeneration, describing ecohydrological processes which the results of the hydrological monitoring programme were only able to hint at. Consequently, it can be concluded that farmers do possess knowledge of the wetland hydrological environment which offers a sound basis for achieving their hydrological and wetland management goals.

In examining how this knowledge is operationalized as real hydrological management practices, it has been suggested that these practices are only partially successful in a way which does not in general reflect farmers' hydrological knowledge. In particular, the hydrological data suggests that the effects of hydrological management are spatially variable and this has consequences on the success of crop cultivation. It is suggested that the principal reason behind the partial success of hydrological management practices is that wetland farmers are forced to contend with a range of constraints. These constraints, for various reasons, prevent hydrological knowledge being fully operationalized as farmers ultimately lack the time or resources to concentrate on hydrological management activities. Although farmers do carry out minor adaptations to their hydrological management practices, there is little evidence to suggest that they are able to cope with the existing range of constraints which limit their hydrological management success.

Although this situation may dominate wetland management throughout the study area, evidence suggests that one key aspect of sustainable hydrological management, i.e. hydrological sustainability, is being achieved. Management practices in their current form are contributing to a state of hydrological sustainability in that where drainage and cultivation is being undertaken, the hydrological characteristics of the wetland remain similar from year to year. Critically, hydrological management does not appear to reduce the capacity of the wetland water table to continue to provide its range of functions and benefits over time.

Whilst hydrological sustainability is facilitating crop production without degradation in the wetlands, it does not appear to sustain a level of crop production which meets the needs of the wetland farmers. Hence it could be argued that output sustainability is not being achieved. The current application of hydrological management practices, under the influence of various constraints, have resulted in a degree of spatial variability and unpredictability in the wetland water table which does affect crop yield, hence hydrological management can not be considered sustainable in this respect.

Finally, the extent to which hydrological management practices are socially sustainable, in that they are able to evolve in response to pressures, is unclear. Although it is suggested that farmers do have some capacity to adapt their management techniques to changing circumstances, at the same time they appear to be unable to cope with various constraints. If this is a result of a lack of innovative potential, opportunity or a poor indigenous communication network, then hydrological management practices are unlikely to evolve and if major environmental or socio-economic changes occur, this may have major implications for both hydrological and crop sustainability.

In conclusion, although it demonstrates some aspects of sustainability, hydrological management in its current form cannot be considered wholly sustainable for the reasons outlined above. Nonetheless, the discussion in this chapter has identified that farmers' hydrological knowledge and the hydrological management practices which are created from this knowledge, do offer the potential for sustainable hydrological management to be attained. Although there are clearly problems with operationalizing their knowledge, wetland farmers do possess the knowledge which can contribute to and be considered key principles behind sustainable hydrological management. These are explored in the next chapter.

Chapter 10

Sustainable Hydrological Management of Wetlands

Introduction

The final chapter draws together the key findings of the research programme. Central to this chapter is the identification of principles for the sustainable hydrological management of wetlands, hence the discussion draws upon the evidence presented in previous chapters, to show that there are clear principles and practices linked to farmers' knowledge and identified through the hydrological research programme, which can contribute to the sustainable hydrological management of wetlands.

Having established the principles of sustainable hydrological management, the chapter goes on to discuss how these principles can be operationalized in the context of the wetland farming constraints outlined in previous chapters. This explores the current and future potential of wetland farmers to develop their wetland knowledge and adapt their hydrological management practices to address these constraints. The long-term sustainability of farmers' hydrological management ultimately rests upon the ability of farmers' knowledge to adapt and evolve to a range of dynamic socio-economic and environmental circumstances. Hence in view of the current state of farmers' adaptive capacity the opportunities for strengthening this are also explored.

The chapter concludes with a discussion of the implications of the research findings for sustainable hydrological management throughout Illubabor zone and in the wider context of Ethiopia.

The principles of sustainable hydrological management

From an examination of hydrological reality, farmers' hydrological knowledge and current hydrological management practices described in previous chapters, the research undertaken in Illubabor drew attention to several aspects of the system of wetland utilization which currently and potentially contribute to sustainable hydrological management with respect to hydrological sustainability, output sustainability and social sustainability. These aspects can, in many respects, be considered key principles for sustainable hydrological management.

The significance of water

Although somewhat obvious, it is worth stressing that the presence of water and a state of hydrological sustainability is the fundamental component of sustainable hydrological management. The seasonal flooding of these wetlands ensures that in their natural state they are able to function as reservoirs of moisture in the landscape, supporting natural vegetation and a range of functions and benefits for the local communities. Even when drained, the presence of residual moisture can ensure successful crop cultivation and, therefore, fulfil the needs of those undertaking drainage and cultivation. Most farmers recognized that over-drainage can have serious consequences on the ability of wetlands to continue to provide their range of benefits, hence maintaining the wetland water balance through hydrological management activities is considered critical in sustaining these benefits (output sustainability based on hydrological sustainability).

Water is, in addition, critical in ensuring hydrological management is socially sustainable. The inherent hydrological variability in each wetland, combined with meteorological variations produce an environment in which farmers have little choice but to adapt, innovate and seek solutions to hydrological management problems. Although their IK network is arguably under-developed, farmers have incorporated a range of adaptive practices and strategies into their wetland farming system (e.g. *cheffe* retention) which facilitate water conservation from year to year in addition to the short-term removal of water from the wetland for crop cultivation. The significance of retaining water in the wetland is highlighted in more detail in the following sections, which focus specifically on its role in wetland regeneration, the design of drainage practices and soil management.

The management of wetland regeneration

The capacity of a wetland to regenerate through the recolonization of *cheffe* is perhaps the single most important mechanism through which hydrological management can be sustainable in terms of sustained wetland benefits and hydrological sustainability. Both the hydrological data and farmers' knowledge suggest that *cheffe* is an important component of the wetland hydrological regime in that its presence promotes and sustains natural wetland characteristics, i.e. a high water table. This capacity is recognized by farmers and, in some cases, *cheffe* recolonization is actively encouraged for its role in regulating the storage and release of water, especially at the headland areas of wetlands. Although *cheffe* is also regarded as a hindrance to the effective drainage of wetlands, the recolonization of *cheffe* vegetation is also recognized by farmers as the means by which the fertility of the wetland can be restored after a period of drainage and cultivation. The decline of the fertility of the wetland is regarded as an inevitable consequence of drainage, although farmers suggest that crop yields tend to be high for several years following the drainage of 'pristine' vegetated sites.

In most cases, however, the abandonment and regeneration of *cheffe* vegetation is not actively part of a farmer's wetland management strategy and involuntary abandonment of individual plots or in some cases the wetland as a whole may occur as a result of farming constraints. In effect, this produces a temporally variable patchwork of land use within a wetland, such as that illustrated at Supe in Figure 10.1 and conceptualized in Figure 10.2. Consequently, it is possible that the situation of hydrological sustainability through mixed land use that appears to exist within the study wetlands is, to a large extent, dependant upon the effects of farming constraints interfering with hydrological management.

Figure 10.1   The variable land use within Supe

Evidence suggests that those wetlands where complete drainage and cultivation has occurred for a number of years (with no sporadic *cheffe* regeneration) have been characterized by degradation in terms of a deterioration in hydrological conditions and the functions and benefits which depend on them. An obvious example is Hurumu wetland but, in addition, Goma Gabriel wetland located outside the study area has a similar management history. The cultivation of Goma Gabriel wetland was initiated by MFM during the early 1990s and after three years of complete drainage and cultivation the wetland was degraded to a dryland environment (Afework Hailu *et al.*, 2000) (Figure 10.3).

The degradation of wetlands to such an extent does, however, appear to be reversible. In both the cases of Hurumu and Goma Gabriel, several years of abandonment have facilitated the recolonization of *cheffe* and the regeneration of natural wetland conditions in a process which would appear to mirror that of the abandonment and regeneration of small wetland plots. The implication, therefore, is that *cheffe* can achieve a regeneration of wetlands and that such wetlands may ultimately regain the characteristics of 'pristine' sites in terms of soil, vegetation

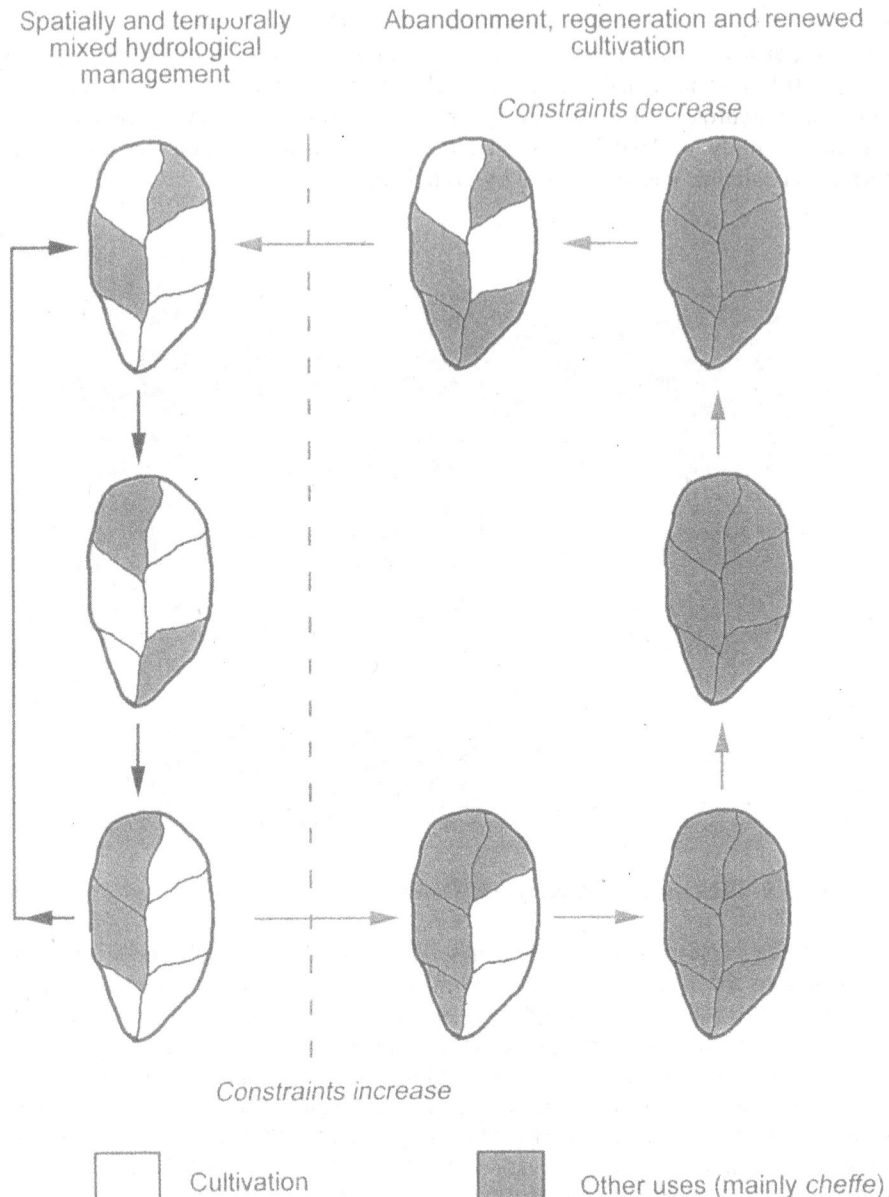

Figure 10.2 A conceptual model of the current situation of cultivation and abandonment in the wetlands

Figure 10.3   The degraded wetland of Goma Gabriel wetland near Bure, pictured after several years of complete cultivation in 1996

and hydrology. These sites when cultivated again, can fulfil the crop requirements of the farmers during the first few critical seasons.

Clearly *cheffe* and the regenerative capacity of wetlands has an important role to play in sustainable hydrological management strategies. Whilst it does contribute to the sustainability of hydrological management at the present time (both in terms of hydrological and output sustainability), an approach to wetland regeneration and *cheffe* management which uses farmers' existing knowledge more explicitly, could assist farmers in fulfilling their hydrological management goals by sustaining crop output and maintaining a water supply throughout the wetland. Two strategies which could potentially balance the need for hydrological sustainability with increased crop output sustainability are proposed in Figure 10.4.

In Figure 10.4, model A proposes that some areas of land within a wetland could remain uncultivated with *cheffe* vegetation. This would reduce some of the hydrological unpredictability inherent in the wetland farming system which stems from sporadic abandonment and, through the pooling of resources, provide the opportunity for farmers to concentrate their hydrological management efforts in specific parts of the wetland. In addition, the uncultivated areas could maintain a supply of *cheffe* (and water) whilst regaining their soil fertility. After several years of operating such a management strategy, sustainable use could be achieved by rotating the land use so that regenerated areas were then cultivated and the previously cultivated areas abandoned. The number of years between each change over is the critical variable. The farmers themselves are knowledgeable with respect to soil fertility and many suggest that crop yields are always high during the first few years of cultivation on 'pristine' wetland.

Maintaining areas of *cheffe* alongside cultivated wetland could present problems with respect to waterlogging at the beginning and end of the cultivation season and the effective drainage of adjacent areas, as is the case in many of the study wetlands. Nonetheless, retaining the same areas of *cheffe* annually for several seasons arguably presents a greater opportunity for farmers to plan and adapt to their hydrological circumstances. Certainly farmers would appear to be aware of the benefits of the approach represented in model A, although whether they would willingly give up cultivating the whole of their wetland plot as part of an overall multiple use strategy is another matter. A more viable alternative could be for farmers to divide their own plots in half so that they cultivated only one half at any time, although such a strategy may be problematic for the owners of smaller wetland plots.

An alternative strategy, which is based upon the evidence that the regeneration of wetlands following periods of drainage does occur, would be one involving the complete cropping of a wetland followed by complete abandonment (Figure 10.4 - model B). Again, the critical factor would be to determine a period of cultivation which did not result in long-term degradation and a period of abandonment which was sufficient to allow the wetlands to regenerate. The potential advantages of this strategy would be higher crop yields during the periods of cultivation and a supply of *cheffe* during fallow periods which could supply a range of natural wetland products and which could also be used sustainably for cattle grazing if access was controlled to prevent overgrazing. In the long-term, successive periods of regeneration and cultivation could sustain the wetland hydrological regime, the fertility of the wetland soil and the range of functions which these wetlands provide in their various states.

An obvious problem would be the periodic harvest failures and food shortages which would require the cultivation of wetlands during fallow periods, although this could be facilitated if several wetlands were involved in a rotational system so that at least one would be under cultivation at any given time. This would also ensure that any sudden demand for *cheffe* was also catered for. Although organization of such a system would also be problematic, within the study area several *kebeles* have designated specific wetlands as a source of *cheffe* (Afework Hailu *et al.*, 2000) suggesting that farmers do appreciate the advantages of resource management at this level. The drainage of pristine sites, however, is very labour intensive, requiring the co-ordination of the whole wetland farming community. The use of several wetlands in rotation would again involve a significant degree of co-operation and organization between farmers who may be reluctant to share the yield from their respective wetlands.

The management of drainage

The management of drainage that is sensitive to environmental conditions can be considered an important principle of sustainable hydrological management. The characteristics of drainage, namely the spacing of drains and their width and depth, determine the water table height in wetland farming plots and subsequently the

Sustainable Hydrological Management of Wetlands 211

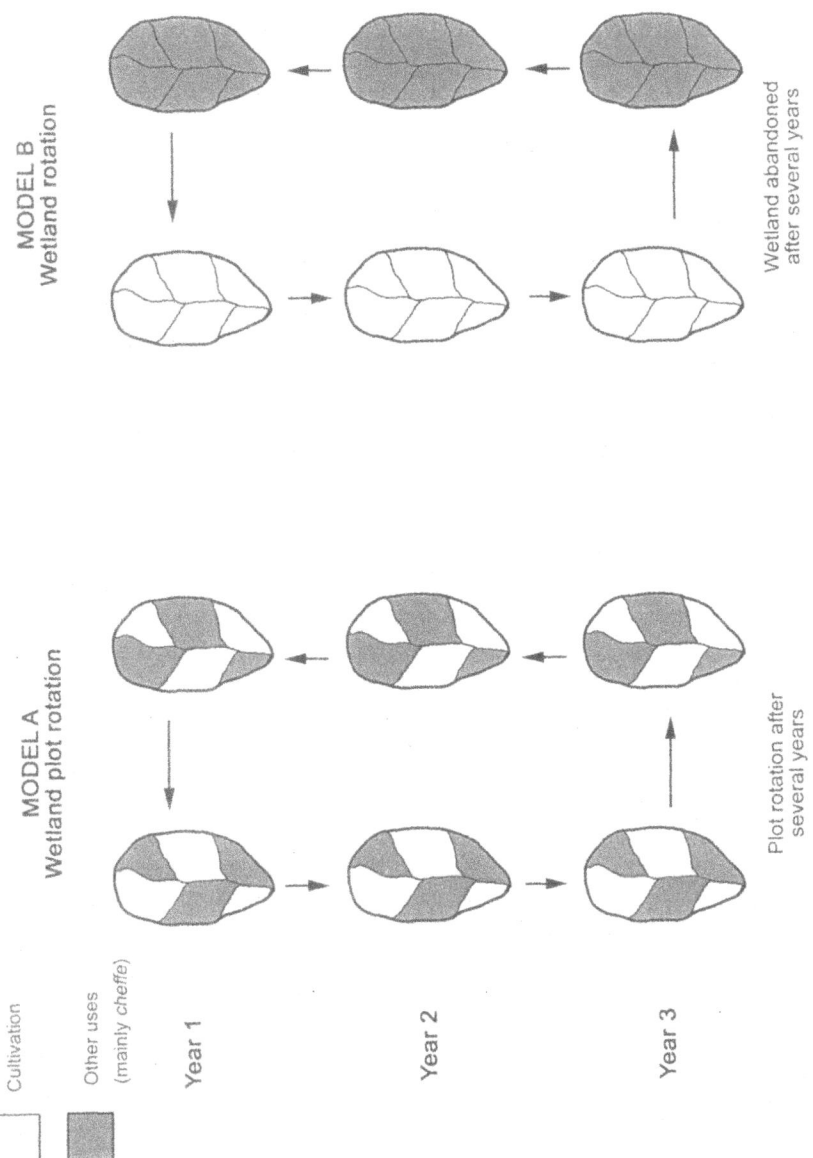

Figure 10.4 Possible strategies for managing wetland regeneration

conditions for crop cultivation. Throughout the study wetlands, however, most drains tends to be less than 1m in width and depth, although the spacing and location of drains varies between sites. These characteristics are, according to farmers, determined by the prevailing hydrological conditions in each wetland and the success of the drainage system in the previous year in terms of crop production.

For the most part, this system of drainage has been successful in that it facilitates crop production year after year and although there is inevitably some disruption to the water table behaviour, there does not appear to be any long-term deterioration in wetland functions and benefits. It could be argued that this situation is a product of the experiences and hydrological knowledge held by wetland farmers, who have developed this system over many decades.

Wetland drainage, however, does not always fulfil the goals of those undertaking it and for this reason it can be regarded as only partially successful. Farmers associate poor crop yields with inadequate drainage, yet faced with a range of other farming constraints they appear to be powerless to make any changes which could increase the efficiency of their drainage practices, despite possessing detailed knowledge of their wetlands' hydrological characteristics. The findings of the research have, nonetheless, highlighted the need for drainage which is suited to the range of soil moisture conditions occurring throughout a wetland, but also carried out in response to seasonal hydrometeorological changes in the wetland.

The spatial variability of the wetland water table also makes drainage problematic in that some areas of each wetland require a greater hydraulic gradient to drain them than others (a result of differences in soil structure). The system of drainage employed by wetland farmers is more suited to wetlands such as Chebere or Tulube, where hydraulic conductivities are high and soil characteristics are more homogeneous. On these sites, fewer drains or shallower drains would be necessary to provide effective drainage. In contrast, the cultivated sites characterized by variable but generally very slow hydraulic conductivities, require closer spaced, deeper drains. If drainage is to be more effective so that farmers can avoid the destructive consequences of waterlogging or over-drainage on their crops, there is a need for farmers to carry out drainage using their precise knowledge of the inherent hydrological variation of their wetland plots.

There are, however, several major drawbacks with such an undertaking. First, the modification of current drainage practices to such an extent would require a massive investment of time and labour, which would not be possible under the present circumstances. Secondly, any increase in the number of drainage ditches could inhibit the use of oxen in the ploughing process. In addition, the co-operation of all the wetland plot owners would be a fundamental requirement, especially as drainage ditches are currently used as a means of demarcating wetland plots. Any changes in the positioning of such drains to suit hydrological conditions rather than land tenure arrangements could precipitate problems within the wetland farming community.

A second critical area of drainage management which contributes to sustainable hydrological management is the continuing practice of drain maintenance throughout the year. In the established farming calendar, drainage is carried out before the growing period and at only two of the study wetlands are modifications made later in the season in response to rainfall. Although farmers regard the spatial variability of the wetland water table as one cause of crop damage, there is no doubt that the duration and timing of rainfall late in the growing season also contributes to the farmers' hydrological management and crop cultivation problems. The unpredictable effects of rainfall could, however, be lessened if farmers were able to extend their drainage management throughout the whole growing season so that drains could be cleared or excavated in response to flooding or alternatively blocked if the rains are delayed.

Again, wetland farmers have the knowledge and experience of typical rainfall patterns and the effects of rainfall anomalies on their hydrological management activities and crops. Carrying out active drainage management throughout the year would, therefore, potentially require the application of existing knowledge, although it is important to ensure that all the wetland farmers possess this knowledge and the capacity to apply it. Such changes to the farming system would again place more demands on the farmer in terms of labour and time (hence there is a question of whether the returns to labour would be sufficient for the farmers).

The management of wetland soils

Careful management of the wetland soils is critical to achieving sustainable hydrological management, as the physical and chemical characteristics of the soils ultimately affect the water table behaviour and the performance of crops. The drainage and cultivation of wetland soils can induce degrading processes such as mineralization, compaction and erosion resulting in a loss of fertility and heterogeneous soil moisture conditions. Although drainage inevitably causes some changes in wetland soil characteristics (Hill, 1976; Barrow, 1991), there is a need to ensure that these are minimized so that the wetlands can sustain the level of crop cultivation required by the farmers.

The destructive effects of drainage and cultivation could be minimized in several ways. First, a more environmentally sensitive approach to drainage could reduce excessive drying out of the wetland soil in specific areas. In addition, the extension of drainage management throughout the season could, as a result of ditch blocking during drought periods, also reduce the drying out and shrinkage of the soil. The key process in soil management is flooding, which is recognized by farmers as a means of improving the structure of the soil (before ploughing) and also improving soil fertility as a result of the input of sediment and the partial decomposition of crop residue (usually maize stalks). The annual blocking of drainage ditches at the end of the cultivation period to maximize the flooding effects of the wet season could constitute a key practice in maintaining the structure and fertility of wetland soils.

In addition to hydrological management practices, the practice of cultivation itself plays an important role in defining the physical and chemical characteristics of the wetland soils. It is critical, therefore, that knowledge of the benefits and impacts of different activities within the system of crop cultivation are shared among the wetland stakeholders, thereby increasing the chances of effective and sustainable soil management. For example, some farmers are aware that the depth of ploughing can affect crop yields if red soils are brought to the surface. In addition, rather than leaving them to decompose in floodwater, some farmers prefer to burn crop residues and plough the ashes into the wetland soil as a means of increasing soil fertility.

There is also a widespread recognition that cattle cause the compaction of wetland soils (Afework Hailu, 1998). Although this is usually associated with overgrazing, cattle are used in activities such as ploughing and weeding, yet no concern was expressed over their impact during these activities. Whilst restricting the use of oxen in the cultivation process would be virtually impossible, it is important that farmers balance their positive contribution to wetland cultivation (ploughing and manuring) with the negative aspects (soil compaction).

There is clearly a role to play for *cheffe* and the regenerative capacity of wetlands in environmentally sensitive soil management. The use of fallow periods for wetland plots or whole wetlands can potentially ensure that wetland soils escape the long-term effects of degradation and, therefore, recover their physical and chemical characteristics to facilitate sustainable hydrological management.

Summary

In summary, several key principles have been identified in the preceding discussions:

1. Maintaining the wetland water balance is fundamental in ensuring that the wetlands do not dry up and thereby lose their functions and benefits which rely on their ecohydrological characteristics.
2. *Cheffe* and the regenerative capacity of wetlands are essential components of sustainable hydrological management on the basis that:
    - *cheffe* promotes and sustains natural hydrological characteristics, it stores water and performs a hydrological regulatory function;
    - the regeneration of *cheffe* is a means by which wetland fertility is restored through sediment trapping and organic matter production;
    - through the regeneration of *cheffe*, degraded wetlands can potentially be restored.
3. Environmentally sensitive management of wetland drainage is fundamental to sustainable hydrological management:
    - drainage should be undertaken with a knowledge of the spatial variability of the wetland water table, so that over-drainage or under-drainage are avoided;
    - drainage practices need to be sensitive to temporal changes in hydrometeorological conditions so that the soil moisture conditions for crop production are optimized throughout the growing season;
    - farmers need to monitor the effects of their drainage (and other hydrological management activities) and make modifications where necessary.

4. Environmentally sensitive management of the wetland soils is essential in providing the optimum conditions for efficient drainage which also minimizes degradation and nutrient loss. This can be achieved through:
   - restricting the impacts of cattle on the wetland soil;
   - promoting soil conservation measures (e.g. ensuring the correct ploughing depth, manuring and increasing the efficiency of cultivation activities);
   - facilitating flooding during the wet season, which improves the soil fertility and structure;
   - ensuring that farmers are aware of the impacts of their own drainage and cultivation practices.

Whilst the adoption of these principles may appear to be a mammoth task for wetland farmers, it should again be stressed that these have been drawn from farmers' existing knowledge and that many farmers are already carrying out such practices in their wetland management strategies. The current state of hydrological sustainability which is being achieved throughout the study wetlands may reflect the farmers' adoption of these principles to the best of their ability under the present circumstances. At the same time, however, these present circumstances are dominated by wetland farming constraints which could be lessened if farmers were to apply the above principles more widely. It is also important to recognize that these principles are inter-related in that the adoption of one may affect the other. For example, although soil management plays an important role in hydrological management, it would to some extent be superfluous if a system of land rotation were adopted, as this would itself have the effect of regenerating soil conditions. These principles should, therefore, be viewed more as a potential set of management options rather than prescriptive practices.

Operationalizing sustainable hydrological management

Whilst the adoption of these principles may increase the potential for sustainable hydrological management, operationalizing them beyond their current level of application may be problematic in that wetland farmers are constrained by various socio-economic and environmental factors. The application of knowledge (and the adoption of such principles) ultimately depends upon the extent to which farmers are able to adapt and cope with these dynamic constraints. Where hydrological knowledge and practices are applied and developed in response to environmental and socio-economic changes, hydrological management can be considered socially sustainable and this is a critical component of sustainable hydrological management. The capacity of farmers to make adaptations, modify and develop their hydrological management practices is itself rooted in their wetland knowledge system.

Clearly there is a need for further, more extensive research on the characteristics and dynamics of wetland knowledge systems within the area. On the basis of the research findings presented here, however, it would appear that the existing IK institutions and networks require some adjustment if they are to guarantee the social sustainability of wetland use and the wider application and adoption of hydrological knowledge.

There is a particular need for indigenous hydrological knowledge to be disseminated to a much greater extent among wetland stakeholders. In terms of innovative capacity (and in the context of other research) the evidence presented suggests that at most, only minor modifications to existing hydrological management practices are made by farmers. Whilst this may reflect the farmers' preoccupation with other farming activities, it is nonetheless important that all the indigenous mechanisms of knowledge acquisition, generation and exchange are strengthened so that farmers can adapt their hydrological management to a range of circumstances, whilst also ensuring it remains sustainable in other ways. Consequently, the maintenance of an effective indigenous communication network and innovative capacity should be regarded as an additional principle for achieving sustainable hydrological management, and this should be a prerequisite to the adoption of those outlined earlier.

The adoption of such a principle, however, may be problematic in that to a large extent it requires farmers to recognize the shortcomings and evaluate the success of their own knowledge system. It also requires the co-operation of farmers within and between wetland communities so that the opportunities for knowledge acquisition and development are increased. Given the current degree of acknowledged communication between farmers, however, it is unlikely that the farmers themselves would initiate a process of assessment and empowerment of their IK. There is, therefore, an opportunity for external agencies, particularly NGOs, to stimulate and participate in such a process of IK resource appraisal (Figure 10.5). Given that such an approach may be problematic in the study area as a result of farmers' extensive experience of top-down government intervention and development agents, it is critical that any relationship between external agencies and farmers is established with care and on an equitable basis. Within their proposed methodology for supporting endogenous development, Haverkort and Hiemstra (1999) suggest that, above all, the ability to listen and empathize are fundamental prerequisites to establishing such a relationship.

Within the framework suggested in Figure 10.5, external agencies would initially play a facilitatory role in bringing together members of the wetland farming community and setting up an environment in which farmers can appraise their own situation so that strengths and weaknesses of the IK network can be identified. The farmers themselves can then address any problems and, through collaboration with external agents, identify ways in which the opportunities for innovation and the exchange of information can be promoted. In particular, this may involve recognizing the role of key individuals within the community who generate or communicate indigenous wetland knowledge. Recent research in the study area, which has established a dialogue between neighbouring communities, has reported positive feedback from farmers on being given the opportunity to exchange knowledge, albeit in a non-indigenous environment (Afework Hailu and Abbot, 1998).

PHASE 1

A - Identifying IK Resources

	EXTERNAL AGENT	COMMUNITY
ROLE	• Facilitation	• Participation • Communication • Discussion

OUTPUT: Identification of IK strengths and weaknesses

B - Developing IK Resources

	EXTERNAL AGENT	COMMUNITY
ROLE	• Facilitation • Logistical assistance • Technical input	• Organization • Co-operation • Discussion

OUTPUT: **Stronger IK Resources**
(communication networks, innovative capacity, adaptive capacity)

PHASE 2

A - Identify current IK applications (*wetland management*)

	EXTERNAL AGENT	WETLAND STAKEHOLDERS
ROLE	• Learning • Assimilate IK	• Participation • Discussion

OUTPUT: **Understanding of the logic behind the management system**

B - Problem identification (*inefficient hydrological management*)

	EXTERNAL AGENT	WETLAND STAKEHOLDERS
ROLE	• Participation • Discussion	• Participation • Discussion

OUTPUT: **Identification of the limitations of management**

C - Interaction of knowledge and expertise

	EXTERNAL AGENT	WETLAND STAKEHOLDERS
ROLE	• Participation • Knowledge exchange • Technical input	• Participation • Knowledge Exchange

OUTPUT: **Application of combined knowledge towards specific problem and development of new management strategies**

Figure 10.5 A framework for empowering IK resources

Once this indigenous capacity has been strengthened, a second phase would involve focusing on the specific issues of sustainable hydrological management. Having established several principles which could contribute to sustainable hydrological management, there is a need for these principles to be discussed by the various wetland communities so that they can be evaluated, adapted or refined to meet the needs or address the constraints of each unique wetland environment. Both IK and scientific enquiry can generate information which is complementary and which can be combined to produce a more detailed picture than would otherwise be available exclusively to one knowledge system. By using a framework such as that proposed by Ryden (1991) both knowledge systems can be used to identify specific problems and interact to discuss and develop the principles outlined earlier.

Whilst Ryden's (1991) 'loop' concept stresses the interaction between external researchers and local communities in the development process, there is a need under the current scenario of wetland management to include a third group of actors in the participatory process. It is important that all the wetland stakeholders are considered in any management initiatives if, in this case, hydrological management is to be sustainable in the long-term and socially sustainable in response to the socio-economic or political demands of government planning. Hence in addressing the principles of sustainable hydrological management, the participatory process should involve not only external researchers and local community members, but also those involved either directly or indirectly in affecting the wetland planning process.

Following a recognition of the principal actors involved in wetland management, the first step towards operationalizing the key principles involves all parties attempting to understand the logic behind current wetland management practices and existing sustainable practices, in terms of the prevailing social, political, economic and environmental circumstances. This can then be followed by a process of identification and recognition of the limitations and constraints of wetland management in its present form and the potential problems facing the application of hydrological knowledge and hydrological management principles.

The final stage sees the interaction of external and indigenous knowledge and expertise, leading to the development of wetland management strategies which incorporate and facilitate the wider application of indigenous hydrological knowledge and the principles for sustainable hydrological management.

Having undergone a process through which their IK network and subsequently their adaptive capacity has been empowered, farmers would, therefore, possess an increased capacity to cope and adapt their hydrological management practices to a range of changing circumstances. This ability to cope, however, remains dependant upon the scale and speed of the changes which take place. It is important to recognize that the principles for sustainable hydrological management are based upon research findings of wetland utilization within a particular part of Illubabor, in which a specific set of environmental, socio-economic and political factors have shaped people's relationship with wetlands. In other parts of Ethiopia, particularly Wellega and Jimma zones where socio-economic and demographic pressures have been arguably much greater, people's interaction with wetlands appears to have been less successful, in that wetland cultivation has resulted in an abundance of degraded and abandoned wetlands (pers. comm. Afework Hailu, 1999).

Past, present and future trends in wetland use

The many inter-related influences upon Illubabor's wetlands which range from political change to localized cattle disease make it difficult to predict any changes which may affect their utilization. This prediction is particularly problematic in that Illubabor has itself undergone relatively rapid socio-economic, environmental and political changes since the 1970s yet, on the basis of this research, this has not resulted in widespread intensive use and the degradation of wetland resources. Whilst increasing coffee production, demands for regional food self sufficiency, demographic growth and the rise of the market economy may have had some impact on wetland use, the extension of wetland agriculture (bringing more wetland areas into use) rather than intensification appears to have been the farmers' response to pressure.

Farmers have the option of draining and cultivating their wetland plots and this depends on a range of factors, the most important of which is the prevailing food security situation. According to most farmers, however, the agricultural use of wetlands peaked during the Haile Selassie era when as tenants they were forced to cultivate wetlands. During successive governments, wetland agriculture appears to have continued in some areas but been abandoned in others and this remains the situation to the present day.

The main implication for the future of wetlands in Illubabor is that they have the potential to survive a range of pressures. They have for the most part, escaped degradation in the past because the cycle of abandonment and cultivation of wetlands (or plots within wetlands) in response to food shortages has facilitated a regenerative process which is typified by a state of hydrological sustainability. The adoption of the principles for sustainable hydrological management outlined here, combined with an empowered IK network, will potentially render wetland management even more robust in the face of increasing pressures. A key concern, however, is if the demand for food increases so much that a higher proportion of wetland area were to be cultivated for an extended period of time, i.e. intensification rather than extensification, thereby preventing the essential regeneration of wetland areas. The result could potentially be the widespread degradation of wetland resources.

Recent indications from the study area in Illubabor are that the preconditions for the intensification of wetland agriculture are prevalent throughout the zone, in that the Ministry of Agriculture have instructed farmers to begin cultivating those wetland plots which are not currently under cultivation, via a 'Wetlands Task Force Committee'. This policy seems to have stemmed from several years of delayed rains which have culminated in a state of food insecurity throughout Illubabor and in particular, Metu and Ale Didu *weredas* (UNDP, 1999). Any subsequent increase in the use of wetlands as a result of this situation presents, in addition to the possibility of wetland degradation, an immediate conflict of interests with those wetland owners who value wetlands as a source of *cheffe* above their agricultural potential. Equally any increase in wetland cultivation is likely to cause problems for those farmers who prefer to graze their cattle in the wetlands or those who rely on wetlands for a range of medicinal plants.

Whilst wetland degradation would be an inevitable outcome from a neo-Malthusian perspective, the above scenario also echoes that described by Boserup (1965) who regards agricultural intensification as the principal means of stimulating technology development to meet the food demands of growing populations. With increasing population pressure, Boserup argues that more labour intensive farming methods are adopted and technological developments are made as farmers aim to increase total levels of food production from agricultural land and maintain long-term food production. Although Boserup's hypothesis has attracted widespread criticism (Grigg, 1979), the case of the Machakos district in Kenya among others (Turner *et al.*, 1993) has been used as an example of population pressure stimulating agricultural development which has not induced widespread environmental degradation (Ondiege, 1992; Tiffen *et al.*, 1994; Mortimore and Tiffen, 1995).

Mortimore and Tiffen (1995) relate how during the 1930s the Machakos district suffered from soil erosion, drought, crop failure and population pressure, with farmers owning less than one hectare of land. In response to these problems, the colonial government initiated a programme of terracing and soil and water conservation measures, which were endogenously developed and modified after independence. Throughout this period it is reported that the area has seen significant changes in terms of technological innovation, food self sufficiency, market growth and soil and water conservation. Rather than population pressure inducing environmental degradation, the resulting land shortages created the conditions for sustainable agricultural intensification and the evolution of the farming system.

At the core of Boserup's theory is the suggestion that 'necessity is the mother of invention' (Findlay and Findlay, 1987), hence it is possible that the increasing use of Illubabor's wetlands for agriculture may catalyze the evolution of wetland knowledge, as farmers (like those in Machakos) are forced to seek new solutions to problems. Evidence from other parts of Ethiopia suggest that indigenous soil and water conservation techniques are well developed in those areas where environmental degradation is a part of everyday life (Kruger *et al.*, 1996; Fetien Abay *et al.*, 1999). In the Harerge highlands for example, increased demographic pressure, frequent land redistribution, unreliable rainfall and declining soil fertility among other factors have led to a decline in agricultural productivity and environmental degradation (Kebede Asrat, *et al.*, 1996). In this context, farmers have become increasingly reliant on a range of indigenous soil and water conservation practices which have been developed over generations. Kebede Asrat *et al.* (1996) report that the construction and modification of soil and stone bunds has occurred incrementally in response to changing local needs and environmental circumstances, facilitating moisture retention in drier areas so that agriculture is successful. Similarly in the northern highlands of Tigray indigenous innovations are widespread throughout an area which has periodically suffered from drought, deforestation and soil erosion (Fetien Abay *et al.*, 1998).

Whilst the intensification of wetland agriculture in response to pressure may induce adaptation and innovation, one major difference is apparent between the conditions described by Boserup and those which exist in Illubabor. In particular, the imminent intensification of wetland agriculture is driven by government policy rather than farmers reacting to their own pressures, something which Boserup's model does not take into account. Furthermore, in an environment of increased state intervention not dissimilar to that of previous governments, in which land remains the property of the state, farmers may have little incentive to invest their resources in adapting and developing their wetland farming system.

It is also important to recognize that there is always likely to be a lag between any population or environmental pressure and farmers' adaptive responses, especially with respect to drainage management. In the intervening period during which farmers come to terms with a new situation, management practices may be inappropriate, unproductive and unsustainable.

The wetlands throughout the study area, therefore, appear to be at a critical point in their history when cultivation may be undertaken on an unprecedented scale, placing increasing demands on the wetland hydrological system. Whilst farmers have demonstrated that their indigenous hydrological management can be sustainable, it is, however, the government who ultimately possess the power to affect wetland use at the present time. The onus is, therefore, on the government to consider the physical and socio-economic relationships between wetland use and upland use, and in addition, the far reaching, long-term consequences of encouraging rapid expansion of wetland cultivation throughout the area. In this respect and in view of the interventionist nature of successive governments in this region, it is unlikely that wetland farmers will ever have the opportunity to operationalize fully their indigenous practices which contribute to sustainable hydrological management.

Conclusions

The circumstances surrounding the use of wetlands in Illubabor are much more complex than those originally envisaged at the outset of the research. Having monitored and identified their hydrological characteristics, a key finding has been the hydrological sustainability of most wetlands throughout the study area and their ability to regenerate following a cessation in agricultural use. This capacity can be attributed to a range of factors, although undoubtedly the inter-relationship between farmers' hydrological knowledge and their ability to apply this knowledge in the face of constraints is the key variable.

The hydrological sustainability of wetlands has also been supported by the evidence to suggest that wetland drainage and cultivation in its current form is not a recent phenomenon. It appears that wetland farming has been carried out in similar manner for generations and critically, socio-economic or demographic changes have had variable effects on different farmers rather than influencing wider trends in wetland use. Hence wetland degradation, in its most apocalyptic sense, has generally been avoided in this part of Illubabor. It does appear, however, that the concerns over increasing wetland use, which precipitated this research, are more relevant at the present time.

The research has also demonstrated that wetlands which can be considered hydrologically sustainable are not necessarily sustainable in terms of their outputs, i.e. there is no certainty that farmers will achieve their wetland management goals. Consequently, if the range of wetland functions and benefits are to be sustained for future generations, it is critical that hydrological sustainability exists alongside the sustainability of wetland outputs, e.g. the production of wetland crops or medicinal plants. The research findings suggest, however, that whilst most farmers do contribute to a state of hydrological sustainability through their hydrological management practices, they are unable to maintain the sustainability of wetland outputs mainly because they are constrained by socio-economic factors which also prevent the wider application of their wetland knowledge.

As in other natural resource management systems throughout the developing world, IK has been identified as playing a fundamental role in wetland management in the study area. Whilst there is some evidence to suggest that farmers can adapt their hydrological knowledge and management practices throughout the accumulation of IK, it is also clear that farmer innovation or the farmer to farmer communication of wetland knowledge is not widespread. If hydrological and output sustainability are to be achieved together, there is a need to empower the IK network so that the adaptive capacity of farmers and the evolution of wetland knowledge is catalyzed through increased opportunities for innovation and communication. Once this is achieved, the potential for a self-supporting wetland management system in which hydrological management is itself sustainable can be maximized.

In wider terms, the research has also made a fundamental contribution to the debate regarding the significance and importance of IK in rural development. Whilst it has been demonstrated that IK can form the basis of a sustainable natural resource management system, the limitations of IK have also been apparent. In particular, within the current context of wetland management it has been suggested that the adaptive capacity and evolution of indigenous wetland knowledge remains relatively slow and that this may precipitate a shift to unsustainable management.

Whilst these findings are consistent with the argument that IK does not necessarily guarantee sustainability during periods of rapid social or environmental change (Farrington and Martin, 1998; McCorkle 1998; Wood, 1991; World Bank, 1998), the research has, in addition, drawn attention to the importance of the wider socio-political context and its associated institutional structures as a more subtle influence on the development of IK and indigenous natural resource management strategies.

It is perhaps unrealistic, therefore, to place such high expectations upon local communities in terms of their IK and its ability to sustain natural resource management systems, particularly when IK appears to be rooted in a socio-political context which may ultimately stultify its own development. Consequently, external or scientific knowledge can play a key role in recognizing and complementing the benefits, limitations and utility of IK, thereby maximizing the potential for sustainable resource management.

Bibliography

Abbink, J. (1995), 'Medicinal and ritual plants of the Ethiopian south-west: an account of recent research', *Indigenous Knowledge and Development Monitor*, 3, 2, pp. 6-9.

Abbot, P.G. (1997), *The use of participatory rural appraisal tools: an introduction*, Wetlands and the Watershed Paper No. 1, EWRP, Metu, Unpublished (mimeo).

Ackermann, E. (1936), 'Dambos in Rhodesia', *Wiss. Veroeff. Leipzig*, 4, pp. 149-157.

Acland, J.D. (1971), *East African Crops*, Longman Group Ltd, London.

Acreman, M.C. and Hollis, G.E. (1996), *Water management and wetlands in sub-saharan Africa*, IUCN, Gland, Switzerland.

Acres, B.D., Blair Rains, A., King, R.B., Lawton, R.M., Mitchell, A.J.B. and Rackham, L.J. (1985), 'African dambos: their distribution, characteristics and use', *Z. Geomorphol., N.F. Supplementband*, 52, pp. 63-86.

Adams, W.M. (1990), *Green development: environment and sustainability in the third world*, Routledge, London.

Adams, W.M. (1992), *Wasting the rain: rivers, people and planning in Africa*, Earthscan, London.

Adams W.M. (1993a), 'Indigenous use of wetlands and sustainable development in west Africa', *The Geographical Journal*, 159, 2, pp. 209-218.

Adams, W.M. (1993b), 'Agriculture, grazing and forestry', in G.E. Hollis, W.M. Adams, and M. Aminu-Kanu (eds), *The Hadejia-Nguru wetlands, environment, economy and sustainable development of a sahelian floodplain wetland*, IUCN, Gland, Switzerland, pp. 89-96.

Adams, W.M. (1993c), 'The wetlands and nature conservation', in G.E. Hollis, W.M. Adams, and M. Aminu-Kanu (eds), *The Hadejia-Nguru wetlands, environment, economy and sustainable development of a sahelian floodplain wetland*, IUCN, Gland, Switzerland, pp. 211-214.

Adams, W.M. (1996), 'Economics and hydrological management of African floodplains', in M.C. Acreman and G.E. Hollis (eds) *Water management and wetlands in sub-saharan Africa*, IUCN, Gland, Switzerland, pp. 21-33.

Adams, W.M. and Hughes, F.M.R. (1990), 'The environmental effects of dam construction in tropical Africa: impacts and planning procedures', *Geoforum* 17, 3, pp. 403-410.

Adamus, P.R and Stockwell, L.T. (1983), *A method for wetland functional assessment: volume 1 - critical review and evaluation concepts*, US Department of Transportation, Federal Highway Administration, Report FHWA-IP-82-83, Washington DC.

AED (Academy for Educational Development) (1988), *Conference on indigenous knowledge systems: implications for agriculture and international development*, Washington DC.

Afework Hailu (1998), *An overview of wetland use in Illubabor*, EWRP, Metu, Unpublished (mimeo).

Afework Hailu and Abbot, P.G. (1998), *Report on the first farmers' workshop, 5th October 1998*, EWRP, Metu, Unpublished (mimeo).

Afework Hailu and Abbot, P.G. (1999), *Annual progress report of the activities of the Ethiopian Wetlands Research Programme field office: July 1998 - May 1999*, EWRP, Metu, Unpublished (mimeo).

Afework Hailu, Abbot, P.G., Dixon, A.B., Wood, A.P., Belay Tegegne and Zerihun Woldu (2000), *Report for objective 4 on sustainable management systems*, EWRP, Metu, Unpublished (mimeo).

Agrawal, A. (1995), 'Dismantling the divide between indigenous and scientific knowledge', *Development and Change*, 26, pp. 413-439.

Alemneh Dejene (1990), *Environment, famine and politics in Ethiopia: A view from the village*, Lynne Reiner Publishers, Boulder and London.

Allan, W. (1949), *Studies in African land usage in Northern Rhodesia*, Rhodes-Livingston Paper No 15.

Amatya, D.M., Chescheir, G.M. and Skaggs, R.W. (1995) 'Hydrologic effects of wetland location and size in an agricultural landscape', in K.L. Campbell, (ed.), *Versatility of wetlands in the agricultural landscape, Proceedings of an AWRA conference, Tampa, Florida, September 17-20 1995*, ASAE, pp. 477-488.

Armstrong, A. (2000), 'Ethical considerations in wetland management', *Phys. Chem. Earth (B)*, 25, 7-8, pp. 641-644.

Asmamaw Legasse (1998), *Soils, soil degradation and soil management systems in and around Chebere wetland, Illubabor, Ethiopia*, Unpublished MSc Thesis, Addis Ababa University, Ethiopia.

Atteh, O.D. (1989), *Indigenous local knowledge as the key to local-level development: possibilities, constraints and planning issues in the context of Africa*, Seminar on reviving local self reliance: challenges for rural/regional development in eastern and southern Africa, Arusha, Tanzania, 21- 24 February 1989.

Azene Bekele-Tesemma (1993), *Useful trees and shrubs for Ethiopia: identification, propagation and management for agricultural and pastoral communities*, Regional Soil Conservation Unit Technical Handbook No 5, SIDA, Nairobi.

Bahru Zewde (1991), *A history of modern Ethiopia 1855 - 1974*, Addis Ababa University Press, Addis Ababa.

Baird, A.J and Ross, M.S. (1992), 'Modelling transient water tables and soil moisture conditions in drained peats', in O.M. Bragg, H.A.P. Ingram and R.A. Robertson (eds), *Peatland ecosystems and man: an impact assessment*, Dept. of Biological Science, University of Dundee, pp. 110-117.

Balek, J. (1977), *Hydrology and water resources in tropical Africa*, Elsevier, Amsterdam.

Balek, J. (1983), *Hydrology and water resources in tropical regions*, Elsevier, Amsterdam.

Balek, J. and Perry, J.E. (1973), 'Hydrology of seasonally inundated African headwater swamps', *Journal of Hydrology*, 19, pp. 227-249.

Barbier, E.B. (1987), 'The concept of sustainable economic development', *Environmental Conservation*, 14, 101-110.

Barbier, E.B. (1993), 'Sustainable use of wetlands - valuing tropical wetland benefits: economic methodologies and applications', *The Geographical Journal*, 159, 1, pp. 22-32.

Barbier, E.B., Acreman, M. and Knowler, D. (1997), Economic valuation of wetlands: a guide for policy makers and planners, Ramsar Convention Bureau, Gland, Switzerland.

Barrow, C.J. (1991), *Land degradation: development and breakdown of terrestrial environments*, Cambridge University Press, Cambridge.

Bartelmus, P. (1986), *Environment and development*, Allen and Unwin, Boston.

Bastian, R.K. and Benforado, J. (1988), 'Water quality functions of wetlands: natural and managed systems', in D.D. Hook, W.H. McKee, H.K. Smith, J. Gregory, V.G. Burrell, M.R. DeVoe, R.E. Sojka, S. Gilbert, R. Banks, L.H. Stolzy, C. Brooks, T.D. Matthews and T.H. Shear (eds) *The ecology and management of wetlands volume 1: ecology of wetlands*, Timber Press, Portland, pp. 87-97.

Bayessa Urgessa (1995), *Natural resources and their conservation*, Report for the Ministry of Coffee and Tea Development, Metu, Illubabor, Ethiopia, Unpublished (mimeo).

Beadle, L.C. (1981), *The inland waters of tropical Africa*, Longman, New York.

Belay Tegegne (1998), *Characteristics of wetland soils of Metu area, Illubabor Zone, Ethiopia*, First year report 1997-1998, EWRP, Metu, Unpublished (mimeo).

Belay Tegegne (1999), *Characteristics and management of wetland soils of Metu wereda, Illubabor Zone*, Final Report, EWRP, Metu, Unpublished (mimeo).

Bell, M. (1979), 'The exploitation of indigenous knowledge or the indigenous exploitation of knowledge: use of what for what?', *IDS Bulletin* 10, 2, pp. 44-50.

Bell, M. and Hotchkiss, P. (1989), 'Political interventions in environmental resource use: dambos in Zimbabwe', *Land Use Policy*, 6, 4, pp. 313-323.

Binns, T. (1994), *Tropical Africa*, Routledge, London.

Binns, T. (ed.) (1995), *People and environment in Africa*, Wiley, Chichester.

Binns, T., Hill, T. and Etienne, N. (1997), 'Learning from the people: participatory rural appraisal, geography and rural development in the 'new' South Africa', *Applied Geography*, 17, 1, 1-9.

Biot, Y. (1989), 'The environmental implications of the green revolution', in *Proceedings of the Symposium on the Sustainability of Agricultural Production Systems in Sub-Saharan Africa, September 4 – 7th 1989*, Aas, Norway, NORAGRIC.

Boserup, E. (1965), *The conditions of agricultural growth*, Allen and Unwin, London.

Bouma, J., Dekker, L.W. and Haans, J.C.F.M. (1980), 'Measurement of depth to water table in a heavy clay soil', *Soil Science*, 130, 5, pp. 264-270.

Brace, S. (1995), 'Participatory rural appraisal - a significant step forward in understanding relationships between poor people and their environments', in T. Binns, (ed.), *People and Environment in Africa*, John Wiley and Sons, Chichester, pp. 39-46.

Brinson, M.M., Kruczynski, W., Lee, L.C., Nutter, W.L, Smith R.D and Whigham, D.F. (1994), 'Developing an approach for assessing the functions of wetlands', in W.J. Mitsch, (ed.) (1994), *Global wetlands: old world and new*, Elsevier Sience, Amsterdam, pp. 615-624.

Brokensha, D, Warren, D & Werner, O (1980) *Indigenous knowledge systems and development*, University Press of America, Lanham.

Brookes, S. and Stoneman, R. (1997), *Conserving bogs: the management handbook*, The Stationery Office, Edinburgh.

Brune, S. (1990), 'The agricultural sector', in S. Pausewang, Fantu Cheru, S. Brune, and Eshetu Chole (eds), *Ethiopia: options for rural development*, Zed Books Ltd, London, pp. 15-29.

Bullock, A. (1992a), 'The role of dambos in determining river flow regimes in Zimbabwe', *Journal of Hydrology*, 134, (1-4), pp. 349-372.

Bullock, A. (1992b), 'Dambo hydrology in southern Africa - review and reassessment', *Journal of Hydrology*, 134, (1-4), 373-396.

Burt, T.P. (1978), *Three simple and low-cost instruments for the measurement of soil moisture properties*, Occasional Paper No. 6, Department of Geography and Geology, Huddersfield Polytechnic, Huddersfield.

Butcher, D.P. and Wood, A.P. (1995), *Sustainable wetland development in highland Illubabor*, Research report no 1, Addis Ababa - Huddersfield British Council Academic link, The University of Huddersfield, Unpublished (mimeo).

Bwathondi, P.O.J. and G.U.J. Mwamsojo, (1993), 'The status of the fishery resource in the wetlands of Tanzania', in G.L. Kamukala and S.A. Crafter, (eds), *Wetlands of Tanzania, Proceedings of a seminar on the wetlands of Tanzania, Ngorogoro, Tanzania, 20-27th November, 1991*, IUCN, Gland, Switzerland, pp. 49-60.

Carter, V. and Novitzki, R.P. (1988), 'Some comments on the relation between groundwater and wetlands', in D.D Hook, W.H. McKee, H.K. Smith, J. Gregory, V.G. Burrell, M.R. DeVoe, R.E. Sojka, S. Gilbert, R. Banks, L.H. Stolzy, C. Brooks, T.D. Matthews and T.H. Shear, T H (eds), *The ecology and management of wetlands volume 1: ecology of wetlands*, Timber Press, Portland, pp. 68-86.

Chabwela, H. (1992a), 'The exploitation of wetland resources by traditional communities in the Kafue Flats and Bangweulu Basin', in E Maltby, P.J. Dugan, and J.C. Lefeuvre, (eds), *Conservation and development: the sustainable use of wetland resources, Proceedings of the Third International Wetlands Conference, Rennes, France, 19 - 23 September 1988*, IUCN, Gland, Switzerland, pp. 31-39.

Chabwela, H. (1992b), 'Wetlands as potential sites for the future of tourism development', in R.C.V. Jeffery, H.N. Chabwela, G. Howard, and P.J. Dugan, (eds), *Managing the wetlands of Kafue Flats and Bangweulu Basin, Proceedings of the WWF-Zambia Wetlands Project Workshop, Musungwa Safari Lodge, Kafue National Park, Zambia, 5 - 7th November 1986*, IUCN, Gland, Switzerland, pp. 71-73.

Chambers, R. (1983), *Rural Development: putting the last first*, Longman, London.

Chambers, R. (1990) *Microenvironments unobserved*, Gatekeeper Series No 22, IIED, London.

Chambers, R. (1992), 'Rural appraisal: rapid, relaxed and participatory', *IDS Discussion Paper No. 311*, IDS, Brighton.

Chambers, R. (1994a), 'The origins and practice of participatory rural appraisal', *World Development*, 22, 7, 953-969.

Chambers, R. (1994b), 'Participatory Rural Appraisal (PRA): Analysis of Experience', *World Development*, 22, 9, pp. 1253-1268.

Chambers, R. (1997) *Whose reality Counts? Putting the first last*, Intermediate Technology Publications, London.

Chambers, R., Pacey, A. and Thrupp, L.A. (eds) (1989), *Farmer first: farmer innovation and agricultural research*, Intermediate Technology Publications, London.

Chiuta, T.M. (1995), 'Indigenous knowledge systems for wetlands conservation: Barotse floodplain, Southern Africa', *IUCN Wetlands Programme Newsletter*, pp. 11, 6-7.

Conway, D. (1999), *Second annual hydrological report*, EWRP, Metu, Unpublished (mimeo).

Conway, G.R. (1997), *The doubly green revolution: food for all in the 21st century*, Penguin, London.

Conway, G.R. and Barbier, E.B. (1990), *After the green revolution*, Earthscan, London.

Cornwall, A. and Fleming, S. (1995), 'Context and complexity: anthropological reflections on PRA', *PLA Notes*, 24, pp. 8-12.

Cornwall, A., Guijt, I. and Welbourn, A. (1994), 'Acknowledging process: methodological challenges for agricultural research and extension', in I. Scoones, and J. Thompson, (eds), *Beyond farmer first: rural people's knowledge, agricultural research and extension practice*, Intermediate Technology Publications, London, pp. 98-117.

Cornwall, A. and Jewkes, R. (1995), 'What is participatory research?', *Soc. Sci. Med.*, 41, 12, pp. 1667-1676.

Cross, M. and Millar, J. (1994), 'Ten years after the famine', *New Scientist*, 144, pp. 1950.

CSA (1997), *Ethiopia statistical abstract*, Central Statistics Authority, Addis Ababa.

Davis, T.J. (1994), *The Ramsar Convention manual: a guide to the convention on wetlands of international importance especially as waterfowl habitat*, Ramsar Convention Bureau, Gland, Switzerland.

DeMerona, B. (1992), 'Fish communities and fishing in a floodplain lake of central amazonia', in E Maltby, P.J. Dugan, and J.C. Lefeuvre, (eds), *Conservation and development: the sustainable use of wetland resources, Proceedings of the Third International Wetlands Conference, Rennes, France, 19 - 23 September 1988*, IUCN, Gland, Switzerland, pp. 165-177.

Denny, P. (1993a), 'Wetlands of Africa: introduction', in D. Whigham, D. Dykjova, and S. Hejny, (eds), *Wetlands of the world: inventory, ecology and management*, Volume 1 Handbook of vegetation science, Kluwer Academic Publishers, Dordreacht, pp. 1-31.

Denny, P. (1993b), 'Eastern Africa', in D. Whigham, D. Dykjova, and S. Hejny, (eds), *Wetlands of the world: inventory, ecology and management*, Volume 1 Handbook of vegetation science, Kluwer Academic Publishers, Dordreacht, pp. 32-46.

Denny, P. (1994), 'Biodiversity and wetlands', *Wetlands Ecology and Management*, 3, 1, pp. 55-61.

Denny, P. and Turyatunga, F. (1992), 'Ugandan wetlands and their management', in E Maltby, P.J. Dugan, and J.C. Lefeuvre, (eds), *Conservation and development: the sustainable use of wetland resources, Proceedings of the Third International Wetlands Conference, Rennes, France, 19 - 23 September 1988*, IUCN, Gland, Switzerland, pp. 77-84.

De Walt, B. (1994), 'Using indigenous knowledge to improve agriculture and natural resource management', *Human Organisation*, 53, 2, pp. 123-131.

Diegues, A.S. (1989), 'The role of cultural diversity and community participation in wetland management in Brazil', in M. Marchand, and H.A. Udo, (eds), *The people's role in wetland management, Proceedings of the International Conference in Leiden, The Netherlands, June 5 – 8th 1989*, Centre for Environmental Studies, Leiden University, pp. 345-351.

Dixon, A.B. (2000), *Indigenous knowledge and the hydrological management of wetlands in Illubabor, Ethiopia*, Unpublished PhD Thesis, The University of Huddersfield.

Dixon, C. (1990), *Rural development in the third world*, Routledge, London.

Dries, I. (1989), 'Development of wetlands in Sierra Leone: farmers rationality opposed to government policy', in M. Marchand, and H.A. Udo, (eds), *The people's role in wetland management, Proceedings of the International Conference in Leiden, The Netherlands, June 5 – 8th 1989*, Centre for Environmental Studies, Leiden University, pp. 833-843.

Dries, I. (1991), 'Development of wetlands in Sierra Leone: farmers' rationality opposed to government policy', *Landscape and Urban Planning*, 20, pp. 223-229.

Drijver, C.A. and Marchand, M. (1985), *Taming the floods: environmental aspects of floodplain development in Africa*, Centre for Environmental Studies, State University of Leiden, Leiden.

Drinkwater, M. (1994) 'Knowledge, consciousness and prejudice: adaptive agricultural research in Zambia', in I. Scoones, and J. Thompson, (eds), *Beyond farmer first: rural people's knowledge, agricultural research and extension practice*, Intermediate Technology Publications, London, 32-41.

Dugan, P.J. (1990), *Wetland conservation: a review of current issues and action*, IUCN, Gland, Switzerland.

Dugan, P.J. (1992), 'Research priorities in wetland science: wetland conservation and agricultural development', in E Maltby, P.J. Dugan, and J.C. Lefeuvre, (eds), *Conservation and development: the sustainable use of wetland resources, Proceedings of the Third International Wetlands Conference, Rennes, France, 19 - 23 September 1988*, IUCN, Gland, Switzerland, pp. 3-10.

Dugan, P.J. (ed.) (1993), *Wetlands in danger*, Mitchell Beazley, London.

EMA (1987), *Geological map of Gore*, Ethiopian Mapping Agency, Addis Ababa.

EMA (1988), *A national atlas of Ethiopia*, Ethiopian Mapping Agency, Addis Ababa.

EMA (1997), Monthly rainfall and temperature records for Metu meteorological station, Ethiopian Meteorological Agency, Addis Ababa Unpublished (mimeo).

Fairhead, J. and Leach, M. (1994), 'Declarations of difference', I. Scoones, and J. Thompson, (eds), *Beyond farmer first: rural people's knowledge, agricultural research and extension practice*, Intermediate Technology Publications, London, pp. 75-79.

FAO (1963), *High dam soil survey project*, Aswan - Deb B C, FAO, Rome.

FAO (1979), *Soil survey investigations for irrigation*, Soils Bulletin No 42, FAO, Rome.

FAO (1995), *Forestry resources and programmes, country information brief*, Food and Agriculture Organisation of the United Nations, Ethiopia.

Farrington, J. and Martin, A. (1988), *Farmer participation in agricultural research: a review of concepts and practices*, Agricultural Administration Unit Occasional Paper 9, ODI, London.

Faulkner, R.D. & Lambert, R.A. (1991), 'The effect of irrigation on dambo hydrology: a case study', *Journal of Hydrology*, 123, pp. 147-161.

FDRE (1995) *Federal Negaret Gazeta of the Federal Democratic Republic of Ethiopia*, Addis Ababa, Ethiopia.

Fetien Abay, Mitku Haile and Waters-Bayer, A. (1998), 'Farmers' innovation in land and water management', *ILEIA Newsletter*, 14, 1, pp. 21-23.

Fetien Abay, Mitku Haile and Waters-Bayer, A. (1999), 'Dynamics in IK: innovation in land husbandry in Ethiopia', *Indigenous Knowledge and Development Monitor*, 7, 2, pp. 14-15.

Fielding, D. and Kirsopp-Reed, K. (1994), 'Indigenous knowledge: the way forwards or backwards in tropical agricultural development?', *Science, Technology and Development*, 12, 2-3, pp. 134-145.

Findlay, A. and Findlay, A. (1987), *Population and development in the third world*, Methuen, London.

Gawler, M. (2002), *Strategies for wise use of wetlands: best practices in participatory management*, Proceedings of a workshop held at the 2nd International Conference on Wetlands and Development, November 1998, Senegal.

Gichuki, F.N. (1992), 'Utilisation and conservation of wetlands: an agricultural drainage perspective', in S.A. Crafter, S.G. Njuguna, and G.W. Howard, (eds), *Wetlands of Kenya, Proceedings of the KWWG Seminar on Wetlands of Kenya, Nairobi, Kenya, 3–5th July 1991*, IUCN, Gland, Switzerland, pp. 147-154.

Gill, G.J. (1991), 'But how does it compare to the real data?', *RRA Notes*, 14, pp. 5-14.

Gilman, K. (1994) *Hydrology and wetland conservation*, John Wiley and Sons, Chichester.

Gilvear, D.J. and McInnes, R.J. (1994), 'Wetland hydrological vulnerability and the use of classification procedures: a Scottish case study', *Journal of Hydrology*, 42, pp. 403-414.

Gluckman, M. (1968) *Economy of the central Barotse plain*, Rhodes-Livingstone Papers, No 7, Second Impression, Lusaka, Rhodes-Livingston Institute.

Gonese, C. (1999), 'The three worlds', *Compas Newsletter*, 1, pp. 20-22.

Gopal, B. (1987), *Water hyacinth*, Aquatic Plant Studies No 1, Elsevier, Amsterdam.

Gopal, B. (1989), 'Wetland mismanagement by keeping people out: two examples from India', in M. Marchand, and H.A. Udo, (eds), *The people's role in wetland management, Proceedings of the International Conference in Leiden, The Netherlands, June 5 – 8th 1989*, Centre for Environmental Studies, Leiden University, pp. 352-359.

Gottlich, Kh. (1977), Öko-hydrologische Untershungen an südwestdeutschen Niedermoorstandorton unter der Einwirkung kulturtechnischer Eingriffe1961-1973, *Schr. R. Kuratorium f. Kulturbauwesen*, H. 30 (Eng. Abstract), Verlag Parey, Hamburg / Berlin.

Government of the Republic of Uganda (1994), *National policy for the conservation and management of wetland resources*, Ministry of Natural Resources, Kampala.

Grenier, L. (1998), *Working with indigenous knowledge: a guide for researchers*, IDRC, Ottawa.

Grigg, D. (1979), 'Ester Boserup's theory of agrarian change: a critical review', *Progress in Human Geography*, 9, pp. 64-84.

Hammersley, M. (1992), *What's wrong with ethnography?*, Routledge, London.

Harper, D. (1992), *Eutrophication of freshwaters: principles, problems and restoration*, Chapman and Hall, London.

Haverkort, B. (1995), 'Agricultural development with a focus on local resources: ILEIA's view on indigenous knowledge', in M.D. Warren, L.J. Slikkerveer and D. Brokensha (eds), *The cultural dimension of development: indigenous knowledge systems*, Intermediate Technology Publications, London.

Haverkort, B. and Hiemstra, W. (eds) (1999), *Food for thought: ancient visions and new experiments of rural people*, ETC / COMPAS, Zed Books, London.

Haverkort B., Hiemstra, W. and van't Hooft, K. (1999), 'In search of the right compass: towards experimentation within farmers' worldviews', *Compas Newsletter*, 1, pp. 4-6.

Heathwaite, L. (1994), 'Eutrophication', *Geography Review*, March 1994.

Heyd, T. (1995), 'Indigenous knowledge, emancipation and alienation', *Knowledge and Policy*, 8, 1, pp. 63-73.

Hill, A.R. (1976), 'The environmental effects of agricultural land drainage', *Journal of Environmental Management*, 4, pp. 251-274.

Hirabayashi, E., Warren, D.M. and Owen, W. (1980), 'That focus on the 'Other 40%': a myth of development', in D. Brokensha, D. Warren, and O. Werner, (eds) (1980), *Indigenous knowledge systems and development*, University Press of America, Lanham.

Hollis, G.E. (1990), 'Environmental impacts of development on wetlands in arid and semi-arid lands', *Hydrological Sciences Journal*, 35, 4, pp. 411-428.

Hollis, G.E., Adams, W.M. and Aminu-Kano, M. (1993), *The Hadejia-Nguru wetlands: environment, economy and sustainable development of a sahelian floodplain*, IUCN, Gland, Switzerland.

Hollis, G.E., Holland, M.M., Maltby, E. and Larson, J.S. (1988), 'Wise use of wetlands', *Nature and Resources*, 24, 1, pp. 2-13.

Horne, A.J. and Goldman, C.R. (1994), *Limnology*, McGraw-Hill, New York.

Howard, G (1991) Ugandan wetlands policy seminar, *IUCN Wetlands Programme Newsletter*, 3, 4-5.

Howard, G.W. (1997), *Report of the technical input of IUCN for the Ethiopian Wetlands Research Programme*, June 1997, EWRP, Metu, Unpublished (mimeo).

Howes, M. (1980), 'The uses of indigenous technical knowledge in development', in D. Brokensha, D. Warren, and O. Werner (eds) (1980), *Indigenous knowledge systems and development*, University Press of America, Lanham, pp. 335-351.

Howes, M. and Chambers, R. (1979), 'Indigenous technical knowledge: analysis, implications and issues', *IDS Bulletin*, 10, 2, pp. 5-11.

Howes, M. and Chambers, R. (1980), 'Indigenous technical knowledge: analysis, implications and issues', in D. Brokensha, D. Warren, and O. Werner (eds) (1980), *Indigenous knowledge systems and development*, University Press of America, Lanham, pp. 323-331.

Hughes, F.M.R. (1996), 'Wetlands', in W.M. Adams, S.A. Goudie and A.R. Orme (eds), *The physical geography of Africa*, Oxford University Press, Oxford, pp. 267-286.

Hughes, J.M.R. (1992). 'Use and abuse of wetlands', in A.M. Mannion and S.R. Bowlby (eds), *Environmental issues in the 1990's*, Wiley, Chichester, pp. 211-226.

Hurni, H. (1988), 'Ecological issues in the creation of famines', Paper presented at the national conference on disaster prevention and preparedness strategy for Ethiopia, Addis Ababa.

IDS (1979), 'Rural development: whose knowledge counts?' *IDS Bulletin* 10, 2, Institute of Development Studies, University of Sussex.

IIRR (1996), *Recording and using indigenous knowledge: a manual*, International Institute of Rural Reconstruction, Silang, Cavite, Philippines.

Immirzi, C.P., Maltby, E. and Clymo, R.S. (1992), *The global status of peatlands and their role in carbon cycling*, A report for Friends of the Earth by the Wetlands Ecosystems Research Group, Department of Geography, University of Exeter, Friends of the Earth, London.

IUCN (1996), *A wetland classification system for East Africa*, Regional Wetlands Biodiversity Group at Mbale, Uganda, May 1996.

Jackson, I.J. (1977), *Climate, water and agriculture in the tropics*, Longman, London.

Jahn, S. (1981), *Traditional water purification in tropical developing countries: existing methods and potential application*, GTZ, Eschborn, Germany.

Jeffery, R.C.V., Chabwela, H.N., Howard, G. and Dugan, P.J. (eds), (1992), *Managing the wetlands of Kafue flats and Bangweulu basin*, Proceedings of the WWF-Zambia Wetlands Project Workshop, Musungwa Safari Lodge, Kafue National Park, Zambia, 5-7th November 1986, IUCN, Gland, Switzerland.

Johnson, A. (1972), 'Individuality and experimentation in traditional agriculture', *Human Ecology*, 1, 2, pp. 149-159.

Juma, C. (1987), *Ecological complexity and agricultural innovation: the use of indigenous genetic resources in Bungoma, Kenya*, IDS Workshop, University of Sussex.

Kebede Asrat, Kederalah Idris and Mesfin Semegn (1996), 'The 'flexbility' of indigenous SWC techniques: a case study of the Harerge highlands, Ethiopia', in C. Reij, I. Scoones and C. Toulmin (eds), *Sustaining the soil: indigenous soil and water conservation in Africa*, Earthscan, London, 156-162.

Kebede Tato (1993), *Evaluation of the environmental components of Menschen für Menschen projects in Ethiopia*, MFM, Addis Ababa, Unpublished (mimeo).

Kefeni Kejela (1991), *The soils of Iri / Hurumu area, Illubabor: their genesis, distribution, classification and agricultural potential*, SCRP, Addis Ababa, Unpublished (mimeo).

Kieft, J. (1999), 'Farmers' decisions and cosmovision', *Compas Newsletter*, 1, pp. 18-19.

Kimble, G.T. (1960), *Tropical Africa*, The Twentieth Century Fund, New York.

Kimmage, K. and Adams, W.M. (1992), 'Wetland agricultural production and river basin development in the Hadejia-Jama'are valley, Nigeria', *The Geographical Journal*, 158, 1, pp. 1-12.

Krhoda, G.O. (1992), 'The hydrology and function of wetlands', in S.A. Crafter, S.G. Njuguna, and G.W. Howard, (eds), *Wetlands of Kenya, Proceedings of the KWWG Seminar on Wetlands of Kenya, Nairobi, Kenya, 3–5th July 1991*, IUCN, Gland, Switzerland, pp. 13-22.

Kulka, T. (1977), 'How far does anything go? Comments on Feyerabend's epistemological anarchism', *Philosophy of the Social Sciences*, 7, pp. 277-287.

Kumelachew Yeshitela (1997), *An ecological study of the forest vegetation of south-western Ethiopia*, Unpublished MSc Thesis, Addis Ababa University, Ethiopia.

Ladejinsky, W (1969) The green revolution in Bihar - the Kosi area: a field trip, *Economic and Political Weekly*, 4, 39.

Lambert, R.A., Hotchkiss, P.F., Roberts, N., Faulkner, R.D., Bell, M. and Windram, A. (1990), 'The use of wetlands (dambos) for micro-scale irrigation in Zimbabwe', *Irrigation and Drainage Systems*, 4, pp. 17-28.

Landon, J.R. (1991), *Booker tropical soil manual*, Longman Scientific and Technical, London.

Lema, A.J. (1996), 'Cultivating the valleys: vinyungu farming in Tanzania', in C. Reij, I. Scoones and C Toulmin (eds), *Sustaining the soil: indigenous soil and water conservation in Africa*, Earthscan, London, pp. 139-144.

Lloyd, J.W., Tellam, J.H., Rukin, N. and Lerner, D. (1993), 'Wetland vulnerability in east Anglia: a possible conceptual framework and generalised approach', *Journal of Environmental Management*, 37, 87- 02.

Lwanga, M. (1999), 'Education and public awareness: a case study for the Ugandan wetlands programme', in Wetlands International, *Community Participation in Wetland Management: Lessons from the Field, Proceedings of workshop 3: Wetlands, local people and development, of the International Conference on Wetlands and Development held in Kuala Lumpur, Malaysia, 9-13th October 1995*, Wetlands International, Dordrecht, pp. 231-235.

McCall, M.K. (1995), *Indigenous technical knowledge in farming Systems in eastern Africa: a bibliography*, Bibliographies in Technology and Social Change, No 9, Iowa State University, Ames.

McCann, J. (1995), *People of the plow: an agricultural history of Ethiopia 1800 - 1990*, University of Wisconsin Press, Wisconsin.

McCorkle, C.M. (1989), 'Towards a knowledge of local knowledge and its importance for agricultural RD and E', *Agriculture and Human Values*, 6, 3, pp. 4-12.

McCorkle, C.M. and McClure, G.D. (1995), 'Farmer know-how and communication for technology transfer: CTTA in Niger', in M.D. Warren, L.J. Slikkerveer and D. Brokensha (1995), *The cultural dimension of development: indigenous knowledge systems*, Intermediate Technology Publications, pp. 323-332.

McEldowney, S., Hardman, D.J. and Waite, S. (1993), *Pollution: ecology and treatment*, Longman Scientific and Technical, London.

McFarlane, M.J. and Whitlow, R. (1990), 'Key factors affecting the initiation and progress of gullying in dambos in parts of Zimbabwe and Malawi', *Land Degradation and Rehabilitation*, 2, pp. 215-235.

McIntosh, R.J. (1983), 'Floodplain geomorphology and the human occupation of the upper inland delta of the Niger', *Geographical Journal*, 149, 2, pp. 182-201.

Mackel, R. (1985), 'Dambos and related landforms in Africa - an example for the ecological approach to tropical geomorphology', *Z. Geomorphol., N.F. Supplementband*, 52, pp. 1-23.

Mafabi, P. and Taylor, A.R.D. (1993), 'The national wetlands programme, Uganda', in T.J. Davis (ed.), *Towards the wise use of wetlands*, Ramsar Convention Bureau, Gland, Switzerland.

Maltby, E. (1986), *Waterlogged wealth: why waste the world's wet places?*, Earthscan, IIED, London.

Maltby, E. and Turner, R.E. (1983), 'Wetlands of the world', *Geographical Magazine*, 55, pp. 12-17.

Maltby, E., Dugan, P.J. and Lefeuvre, J.C. (1992), *Conservation and development: the sustainable use of wetland resources*, Proceedings of the Third International Wetlands Conference, Rennes, France, 19-23rd September 1988, IUCN, Gland, Switzerland.

Maltby, E., Hogan, D.V., Immirzi, C.P., Tellam, J.H. and van der Peijl, M.J. (1994), 'Building a new approach to the investigation and assessment of wetland ecosystem functioning', in W.L. Mitsch (ed.), *Global wetlands: old world and new*, Elsevier, Amsterdam, pp. 637-658.

Masundire, H. (1996), 'The effects of the Kariba dam and its management on the people and ecology of the Zambezi river', in M.C. Acreman and G.E. Hollis (eds), *Water management and wetlands in sub-saharan Africa*, IUCN, Gland, Switzerland, pp. 107-118.

Matiza, T. and Crafter, S. (1994), *Wetlands ecology and priorities for conservation in Zimbabwe*, Proceedings of a Seminar on Wetlands of Zimbabwe, Harare, 13-15th January, 1992, IUCN, Gland, Switzerland.

Matthews, G.V.T. (1993), *The Ramsar convention on wetlands: its history and development*, Ramsar Convention Bureau, Gland, Switzerland.

Mermet, L. (1989), 'Participation, strategies and ethics: roles of people in wetland management', in M. Marchand and H.A. Udo (eds), *The people's role in wetland management, Proceedings of the International Conference in Leiden, The Netherlands, June 5 – 8th 1989*, Centre for Environmental Studies, Leiden University, pp. 92-100.

MFM (1995), *1995 Eco-development project work plan*, MFM, Illubabor, Unpublished (mimeo).

Mihayo, J.M. (1993), 'Water supply from wetlands in Tanzania', in G.L. Kamukala and S.A. Crafter (eds) (1993) *Wetlands of Tanzania, Proceedings of a Seminar on the Wetlands of Tanzania. Morogoro, Tanzania, 20-27th November 1991.* IUCN, Gland, Switzerland, pp. 67-72.

Millar, D. (1993), 'Farmer experimentation and the cosmovision paradigm', in W. deBoef, K. Amanor, K. Wellard and A. Bebbington (eds), *Cultivating knowledge: genetic diversity, farmer experimentation and crop research*, Intermediate Technology Publications, London, pp. 44-50.

Ministry of Agriculture (1998), *Annual report of Illubabor zone, Ministry of Agriculture*, Metu, Illubabor, Ethiopia, Unpublished (mimeo).

Ministry of Agriculture (1999), *Wereda map of Illubabor*, Ministry of Agriculture, Metu, Illubabor, Ethiopia, Unpublished (mimeo).

Mitsch, W.J. and Gosselink, J.G. (1986), *Wetlands*, Van Nostrand Reinhold, New York.

Mitsch, W.J., Mitsch, R.H. and Turner, R.E. (1994), 'Wetlands of the old and new worlds: ecology and management', in W.L. Mitsch (ed.), *Global wetlands: old world and new*, Elsevier, Amsterdam, pp. 3-56.

Moris, J. (1991), *Extension alternatives in tropical Africa*, Overseas Development Institute, London.

Mortimore, M. and Tiffen, M. (1995), 'Population and environment in time perspective: the Machakos story', in T. Binns (ed.) *People and environment in Africa*, Wiley, Chichester, pp. 69-89.

Mundy, P.A. and Compton, J.L. (1995), 'Indigenous communication and indigenous knowledge', in M.D. Warren, L.J. Slikkerveer and D. Brokensha (eds) (1995), *The cultural dimension of development: indigenous knowledge systems*, Intermediate Technology Publications, pp.112-123.

Muthuri, F. (1993), 'Plant products from freshwater wetlands', in S.A. Crafter, S.G. Njuguna and G.W. Howard (eds), *Wetlands of Kenya, Proceedings of the KWWG Seminar on Wetlands of Kenya, Nairobi, Kenya, 3–5th July 1991*, IUCN, Gland, Switzerland, pp. 109-113.

NDPC (1988), *Report on the guidelines for wetland cultivation*, Natural Disaster Prevention Committee, Ethiopia, Unpublished (mimeo).

Naucke, W., Heathwaite, A.L., Eggelsmann, R. and Schuch, M. (1993) 'Mire chemistry', in A.L. Heathwaite and Kh. Göttlich (eds), *Mires: process, exploitation and conservation*, Wiley, Chichester, pp. 263-310.

Nkwi, P. and Toornstra, F. (1989), 'Overview paper: traditional uses, risks and potentials for wise use', in M. Marchand and H.A. Udo (eds), *The people's role in wetland management, Proceedings of the International Conference in Leiden, The Netherlands, June 5 – 8th 1989*, Centre for Environmental Studies, Leiden University, pp. 339-344.

Noble, R. (1996), 'Wetland management in Malawi as a focal point for ecologically sound agriculture', *ILEA Newsletter*, 12, 2, pp. 9.

Oakley, P. and Drijver C.A. (1989) 'Overview paper: is there a science of people's participation in wetland management?', in M. Marchand and H.A. Udo (eds), *The people's role in wetland management, Proceedings of the International Conference in Leiden, The Netherlands, June 5 – 8th 1989*, Centre for Environmental Studies, Leiden University, pp. 773-780.

Odum, E.P. (1979), 'The value of wetlands: a hierarchical approach', in P.E. Greeson, P.E. Clark and J.R. Clark (1979) *Wetland functions and values: the state of our understanding*, AWRA.

OECD (1996), *Guidelines for aid agencies for improved conservation and sustainable use of tropical and sub-tropical wetlands*, Guidelines on aid and development No. 9, OECD.

Omari, C.K. (1993), 'Social and cultural values of wetlands in Tanzania', in G.L. Kamukala, and S.A. Crafter (eds) (1993), *Wetlands of Tanzania, Proceedings of a Seminar on the Wetlands of Tanzania. Morogoro, Tanzania, 20-27th November 1991.* IUCN, Gland, Switzerland, pp. 95-102.

Ondiege, P.O. (1992), 'Local coping strategies in Machakos district, Kenya', in D.R. Fraser Taylor and F. Mackenzie (eds), *Development from within: survival in rural Africa*, Routledge, London, pp. 125-147.

Pankhurst, A. (1990), 'Resettlement, policy and practice', in S. Pausewang, Fantu Cheru, S. Brune and Eshetu Chole (eds), *Ethiopia: options for rural development*, Zed Books Ltd, London, pp. 121-134.

Pausewang, S. (1990), '"Meret le arrashu" Land tenure and access to land: a socio-historical overview', in S. Pausewang, Fantu Cheru, S. Brune and Eshetu Chole (eds), *Ethiopia: options for rural development*, Zed Books Ltd, London, pp. 38-48.

Pearce, D.W. (1988), 'The sustainable use of natural resources in developing countries', in R.K. Turner (ed.) (1988), *Sustainable environmental management principles and practice*, Westview Press, Boulder.

Pearce, D.W. and Turner, R.K. (1990) Economics of natural resources and the environment, Harvester Wheatsheaf, Hemel Hempstead.

Pearse, A. (1980), *Seeds of plenty, seeds of want: social and economic implications of the green revolution*, Clarendon Press, Oxford.

Pretty, J., Guijt, I., Scoones, I. and Thompson, J. (1992), 'Regenerating agriculture: the agroecology of low external-input and community based development', in J. Holmberg (ed.), *Policies for a small planet*, Earthscan, London, pp. 91-123.

Pretty, J. and Shah, P. (1999), 'Soil and water conservation: a brief history of coercion and control', in F. Hinchcliffe, J. Thompson, J. Pretty and P. Shah (eds), *Fertile ground: the impacts of participatory watershed management*, IT Publications, London, pp. 1-12.

Price, M. (1996), *Introducing groundwater*, Chapman and Hall, London.

Rajasekaran, B. (1993), *A framework for incorporating indigenous knowledge systems into agricultural research, extension and NGOs for sustainable agricultural development*, Studies in technology and Social Change No 21, Ames, IA: Technology and Social Change Programme, Iowa State University.

Rajasekaran, B., Warren, D.M. and Babu, S.C. (1991), 'Indigenous natural resource management systems for sustainable agricultural development - a global perspective', *Journal of International Development*, 3, 4, pp. 387-401.

Ramsar Bureau (1971), *Proceedings of the international conference on conservation of wetlands and waterfowl, Ramsar, Iran, 30th January - 3rd February 1971*, International Wildfowl Research Bureau, Slimbridge.

Ramsar Bureau (1999), *Press release, 7th meeting of the conference of the contracting parties to the convention on wetlands (Ramsar, Iran 1971)*, San Jose, Costa Rica, 10-18th May 1999.

Rast, W. and Holland, M. (1998), 'Eutrophication of lakes and reservoirs: a framework for making management decisions', *Ambio*, 17, 1, pp. 2-12.

Reeve, R.C. (1986), 'Water potential piezometry', in A. Klute (ed.), *Methods of soil analysis part 1: physical and mineralogical methods*, Agronomy Monograph no 9, 545-61, American Society of Agronomy, Soil Science Society of America, Madison, Wisconsin.

Reid, C., (1995), *Sustainable development: an introductory guide*, Earthscan, London.

Reij, C., Scoones, I. and Toulmin, C. (1996) *Sustaining the soil, indigenous soil and water conservation in Africa*, Earthscan, London.

Reijntjes, C., Haverkort, B. and Waters-Bayer, A. (1992), *Farming for the future: an introduction to low-external-input and sustainable agriculture*, ILEIA / Macmillan, London.

Rhoades, R.E. and Bebbington, A. (1995), 'Farmers who experiment: an untapped resource for agricultural research and development', in M.D. Warren, L.J. Slikkerveer and D. Brokensha (1995), *The cultural dimension of development: indigenous knowledge systems*, Intermediate Technology Publications, pp. 296-307.

Richards, P. (1980), 'Community environmental knowledge in African rural development', in D. Brokensha, D. Warren and O. Werner (eds), (1980), *Indigenous knowledge systems and development*, University Press of America, Lanham, pp. 181-194.

Richards, P. (1985), *Indigenous agricultural revolution: ecology and food production in west Africa*, Hutchinson, London.

Richards, P. (1986), *Coping with hunger: hazard and experimentation in a west African rice-farming system*, Allen and Unwin, London.

Richards, P. (1989), 'Agriculture as a performance', in R. Chambers, A. Pacey and L.A. Thrupp (eds), *Farmer first: farmer innovation and agricultural research*, Intermediate Technology Publications, London, pp. 39-43.

Richards, P. (1995), 'Participatory rural appraisal: a quick and dirty critique', *PLA Notes*, 24, pp. 13-16.

Roberts, N. (1988), 'Dambos in development: management of a fragile ecological resource', *Journal of Biogeography*, 15, pp. 141-148.

Roberts, N. and Lambert, R. (1990), 'Degradation of dambo soils and peasant agriculture in Zimbabwe', in J. Boardman, I.D.L. Foster and J.A. Dearing (eds) *Soil erosion on agricultural land*, John Wiley and Sons, pp. 537-558.

Rocheleau, D.E.M. (1987), *The user perspective and the agro-forestry research and action agenda*, Paper presented at workshop on 'Farmers and agricultural research: complimentary methods', IDS, University of Sussex, 28 - 31 July 1987.

Roggeri, H. (1998), *Tropical freshwater wetlands: a guide to current knowledge and sustainable management*, Kluwer Academic Publishers, Dordrecht.

Röling, N. (1994), 'Facilitating sustainable agriculture: turning policy models upside down', in I. Scoones and J. Thompson (eds), *Beyond farmer first: rural people's knowledge, agricultural research and extension practice*, Intermediate Technology Publications, London, pp. 245-248.

Roscoe, J. (1923), *The Banyankole*, Cambridge University Press, Cambridge.

Rowell, D.L. (1994), *Soil science: methods and applications*, Longman Scientific and Technical, Harlow.

Ryden, P. (1991), 'The IUCN approach in the Sahel', in A.P. Wood and P. Ryden (eds), *The IUCN Sahel Studies 1991*, IUCN, Gland, Switzerland, pp. 1-5.

Sachs, W. (1992), *The development dictionary*, Zed Books, London.

Sanyanga, R.A. (1994), 'Tourism and wetlands management in Zimbabwe with special reference to the Zambezi river system', in T. Matiza and S.A. Crafter (eds), *Wetlands ecology and priorities for conservation in Zimbabwe, Proceedings of a seminar on the wetlands of Zimbabwe, Harare, 13-15th January 1992*, IUCN, Gland, Switzerland, pp. 79-85.

Schultz, T.W. (1964), *Transforming traditional agriculture*, Yale University Press, New Haven.

Schumacher, E.F. (1973), *Small is beautiful: economics as if people mattered*, Blond and Briggs, London.

Scoones, I. (1991), *Wetlands in drylands: the agro-ecology of savanna systems in Africa*, IIED Drylands programme, London.

Scoones, I. (1995), 'PRA and anthropology: challenges and dilemmas', *PLA Notes*, 24, pp. 17-20.

Scoones, I. and Thompson, J. (eds) (1994), *Beyond farmer first: rural people's knowledge, agricultural research and extension practice*, Intermediate Technology Publications, London.

Shiva, V. (1988), *Staying alive: women, ecology and development*, Zed Books, London.

Siegel, D.I. (1988), 'A review of the recharge-discharge function of wetlands', in D.D. Hook, W.H. McKee, H.K. Smith, J. Gregory, V.G. Burrell, M.R. DeVoe, R.E. Sojka, S. Gilbert, R. Banks, L.H. Stolzy, C. Brooks, T.D. Matthews and T.H. Shear (eds), *The ecology and management of wetlands volume 1: ecology of wetlands*, Timber Press, Portland, pp. 59-67.

Sietchiping, R. (1998), 'Peasant ingenuity and innovation in the face of crisis', *Indigenous Knowledge and Development Monitor*, 6, 3, p. 6.

Silvius, M.J., Oneka, M and Verhagen, A. (2000), 'Wetlands: lifeline for people at the edge', *Phys. Chem. Earth (B)*, 25, 7-8, pp. 645-652.

Smith-Carrington, A.K. (1983), *Hydrological bulletin for the Bua catchment: water resource Unit No. 5*, Report of the Groundwater Section, Department of Lands, Valuation and Water, Lilongwe, Malawi.

Soerjani, M. (1992), 'Utilisation of Wetland Plant Resources by Rice Farmers in Indonesia', in E. Maltby, P.J. Dugan and J.C. Lefeuvre (eds), *Conservation and development: the sustainable use of wetland resources, Proceedings of the Third International Wetlands Conference, Rennes, France, 19-23rd September 1988*, IUCN, Gland, Switzerland, pp. 21-29.

Solomon Abate (1994), *Land use dynamics, Soil degradation and potential for sustainable use in Metu Area, Illubabor Region, Ethiopia*, African Studies Series A13, University of Berne, Switzerland.

Solomon Mulugeta (1999), *Socio-economic determination of wetland use in Illubabor: the case of Dizi, Geba, Kawo and Chatu, and Tulube*, Second Year Report for the Ethiopian Wetland Research Programme, Metu, Ethiopia, Unpublished (mimeo).

Solomon Tekalegn (1998), *Soils and soil management practices in Tulube catchment, Illubabor highlands, Ethiopia*, Unpublished MSc Thesis, University of Addis Ababa, Ethiopia.

Stewart, S. (1995), *Participatory rural appraisal: abstracts of sources, an annotated bibliography*, Development Bibliography 11, Institute of Development Studies, University of Sussex, UK.

Summerfield, M.A. (1991), *Global geomorphology*, Longmans, Harlow.

Swift, J.J. (1979), 'Notes on traditional knowledge, modern knowledge and rural development', *IDS Bulletin*, 10, 2, pp. 41-43.

TAMS-ULG (1996), *Baro-Akobo river basin integrated development master plan project: final report*, Report for the Ministry of Water Resources, Addis Ababa, Ethiopia.

Tafesse Asres (1996), *Agro-ecological zones of south-west Ethiopia*, Unpublished MSc Thesis, University of Trier.

Tagoe, C. (1983), *A tentative review of agriculture in the highlands*, FAO / Ministry of Agriculture joint project, EHRS wp 1.

Taye Mengistae (1990), 'Urban-rural relations in agrarian change: an historical overview', in S. Pausewang, Fantu Cheru, S. Brune and Eshetu Chole (eds), *Ethiopia: options for rural development*, Zed Books Ltd, London, pp. 30-37.

Tegegne Sishaw (1998), *Socio-economic aspects of land use change and wetland utilisation in two communities in Illubabor zone*, Unpublished MSc Thesis, University of Addis Ababa.

Thole, L.S. and Dodman, T. (1996), 'Traditional and modern approaches to community wetland management in Zambia', in Wetlands International, *Community Participation in Wetland Management: Lessons from the Field, Proceedings of workshop 3: Wetlands, local people and development, of the International Conference on Wetlands and Development held in Kuala Lumpur, Malaysia, 9-13th October 1995*, Wetlands International, pp. 202-213.

Thomas, D. (1996), 'Water management and rural development in the Hadejia-Nguru wetlands, north-east Nigeria', in M.C. Acreman and G.E. Hollis (eds), *Water management and wetlands in Sub-Saharan Africa*, IUCN, Gland, Switzerland, pp. 91-100.

Thompson, J.R. (1996), 'Africa's floodplains: a hydrological overview', in M.C. Acreman and G.E. Hollis (eds), *Water management and wetlands in Sub-Saharan Africa*, IUCN, Gland, Switzerland, pp. 5-20.

Thrupp, L. (1989a), 'Legitimising local knowledge: "scientized packages" or empowerment for third world people', in D. Warren, J. Slikkerveer and S. Titilola (eds), *Indigenous knowledge systems: implications for agriculture and international development*, Studies in Technology and Social Change, No 11. Ames, Iowa: Technology and Social Change Program, Iowa State University.

Thrupp, L. (1989b), 'Legitimising local knowledge: from displacement to empowerment for third world people', *Agriculture and Human Values*, 6, 3, pp. 13-24.

Tiffen, M., Mortimore, M. and Gichuki, F. (1994), *More people, less erosion: environmental recovery in Kenya*, John Wiley, Chichester.

Timberlake, L. (1988), *Africa in Crisis*, Earthscan, London.

Trafford, B.D. (1983), 'Drainage design', in A.J. Thomasson and D. Mackney (eds), *Soils and field drainage*, Soil Survey of England and Wales, Harpenden, pp. 5-17.

Trapnell, C.G. and Clothier, J.N. (1937), *The soils, vegetation and agriculture systems of north-western Rhodesia*, Government of Northern Rhodesia, Lusaka.

Tudorancea, C., Zinabu Gebre Mariam and Elias Dadebo (1999), 'Limnology in Ethiopia', in R.G. Wetzel and B. Gopal (eds), *Limnology in developing countries*, International Scientific Publications, New Delhi, pp. 63-118.

Tukahirwa, E.M. (1989), 'Land Use Pressures and Wetlands in Kigezi, South-Western Uganda: A case for rehabilitation and management policy', in M. Marchand and H.A. Udo (eds), *The people's role in wetland management, Proceedings of the International Conference in Leiden, The Netherlands, June 5-8th 1989*, Centre for Environmental Studies, Leiden University, pp. 844-852.

Turner, B. (1986), 'The importance of dambos in African agriculture', *Land Use Policy*, 3, 4, pp. 343-347.

Turner, B. (1994), 'Small-scale irrigation in developing countries', *Land Use Policy*, 11, 4, pp. 251-261.

Turner, B.L., Hyden, G. and Kates, R. (eds) (1993), *Population growth and agricultural change in Africa*, University Press of Florida, Gainesville.

Turner, R.K. (1988), 'Wetland conservation: economics and ethics', in D. Collard, D.W. Pearce and D. Ulph (eds), *Economics, growth and sustainable environments: essays in memory of Richard Lecomber*, MacMillan, London, pp. 121-151.

Turner, R.K. (1991), 'Economics and wetland management', *Ambio*, 20, 2, pp. 59-63.

UNDP (1999), *Prosperity fades: Jimma and Illubabor zones of Oromia region, Emergency Unit for Ethiopia- assessment mission*, EUE, Ethiopia.

VanBeers, W.F.J. (1963), *The auger-hole method: a field measurement of the hydraulic conductivity of soil below the water table*, International Institute for Land Reclamation and Improvement: Bulletin No 1, Wageningen.

Wang, G. (1982), *Indigenous communication systems in research and development*, Paper presented at the conference on knowledge utilisation: theory and methodology, 25-30th April 1982, Honolulu, HI, East-West Centre.

Ward, R.C. and Robinson, M. (1990), *Principles of hydrology*, McGraw Hill, London.

Warren, D.M. (1989), 'Linking scientific and indigenous agricultural systems', in J. Lin Compton (ed.), *The transformation of international agricultural research and development*, Lynne Rienner, Boulder, pp .153-170.

Warren, D.M. (1991), 'Using indigenous knowledge in agricultural development', *World Bank Discussion Paper 127*, World Bank, Washington DC.

Warren, D.M. (1996), Comments on article by Dr Agrawal, *Indigenous Knowledge and Development Monitor*, 4, 1.

Wetlands International (1998), 'Wetlands: a source of life: conclusions of the 2nd international conference on wetlands and development', 10-14th November 1998, Dakar, Senegal.

Wheeler, B.D. (1995), 'Introduction: restoration and wetlands', in B.D. Wheeler, S.C. Shaw, W.J. Fojt, and R.A. Robertson (eds), *Restoration of temperate wetlands*, Wiley, Chichester, pp. 1-18.

Wheeler, B.D. and Shaw, S.C. (1995), *Restoration of damaged peatlands*, Department of the Environment, HMSO, London.

Whitlow, J.R. (1983), 'Vlei cultivation in Zimbabwe', *Zimbabwe Agricultural Journal*, 3, pp. 125-135.

Whitlow, J.R. (1985), 'Dambos in Zimbabwe: a review', *Annals of Geomorphology*, 52, pp. 115-146.

Winarto, Y.T. (1994), 'Encouraging knowledge exchange: integrated pest management in Indonesia', in I. Scoones, and J. Thompson (eds), *Beyond farmer first: rural people's knowledge, agricultural research and extension practice*, Intermediate Technology Publications, London, pp. 150-155.

Wolf, E.C. (1986), *Beyond the green revolution: new approaches for third world agriculture*, World Watch Institute, Washington DC.

Wood, A.P. (1977), *Resettlement in Illubabor province, Ethiopia*, Unpublished PhD thesis, University of Liverpool.

Wood, A.P. (1985), 'A century of development measures and population redistribution along the Upper Zambezi', in J.I. Clarke, M. Khogali, and L.A. Kosinski (eds), *Population and development projects in Africa*, Cambridge University Press, Cambridge, pp. 163-175.

Wood, A.P. (1990), 'Natural resource management and rural development in Ethiopia', in S. Pausewang, Fantu Cheru, S. Brune and Eshetu Chole (eds), *Ethiopia: options for rural development*, Zed Books Ltd, London, pp. 187-198.

Wood, A.P. (1991), 'Beyond land use planning: towards a participatory approach to land use planning and husbandry development for improved natural resource use', in A.P. Wood and P. Ryden (eds), *The IUCN Sahel Studies 1991*, IUCN, Gland, Switzerland, pp. 113-126.

Wood, A.P. (1995), *Sustainable wetland management in Illubabor zone, south-west Ethiopia*, Research proposal submitted to the European Union, Budget Line B7-5040 Environment in Developing Countries, The University of Huddersfield, Unpublished (Mimeo).

Wood, A.P. (1996), 'Wetland drainage and management in south-west Ethiopia: some environmental experiences of an NGO', in A. Reenburg, H.S. Marcusen and I. Nielsen (eds), *The Sahel Workshop 1996*, Institute of Geography, University of Copenhagen, Copenhagen.

Wood, A.P. and Stahl, M. (1989), *Ethiopia national conservation strategy: Phase One Report*, IUCN, Gland, Switzerland.

Woodhouse, P., Bernstein, H. and Hulme, D. (2000), *African enclosures? The social dynamics of wetlands in drylands*, James Currey, Oxford.

World Bank (1998), *Indigenous knowledge for development: a framework for action*, Knowledge and Learning Centre, Africa Region, The World Bank, Washington DC.

Yizelkal Fantahun (1998), *Characteristics and classification of wetland soils in Dizi, Hurumu and Wangeneye; South-Western Highlands of Ethiopia*, Unpublished MSc Thesis, University of Addis Ababa, Ethiopia.

Zerihun Woldu (1998), *Annual biodiversity report 1997 - 1998*, EWRP, Metu, Unpublished (mimeo).

Zimmerer, K.S. (1991), 'Wetland production and smallholder persistence: agricultural change in a highland Peruvian region', *Annals of the Association of American Geographers*, 81, 3, pp. 443-463.

Zwahlen, R. (1996), 'Traditional methods: a guarantee for sustainability?', *Indigenous Knowledge and Development Monitor*, 4, 3, pp. 4-5.

Index

adaptation 37-39, 47, 145, 149, 176, 178-179, 202, 221
Amhara peoples 60, 168

Baro, River 53, 61, 105
biodiversity 3, 17, 22, 27, 29, 48, 61, 71
bulk density, of soil 103, 106, 111-112, 130, 145
burning, in wetlands 46, 104, 145, 160, 163

cattle 105, 109
Cheffe (*Cyperus latifolius*) 62, 69, 89, 103-118, 147, 155, 158-159, 165-170, 176-178, 188, 194, 197, 199, 202, 206-210, 214, 220
cluster analysis 79, 98, 128, 130
Coffee 53, 57, 61, 65, 108-111, 114-115, 152, 165, 169, 171, 201, 219
Commercialization 4
Communication 37- 40, 92, 149, 173, 175, 177, 204, 216-217, 222
compaction, of soil 134, 145, 147, 165, 194, 213, 214
constraints, affecting wetland use 6, 158, 160-163, 168-172, 179, 199, 201, 202-207, 212, 215, 218, 221
Cosmovision paradigm 38, 44, 48

dambos 24, 25, 26, 46, 132, 134
Derg 59-60, 65-67, 106-108, 112, 114, 117, 169, 172, 177
developing countries 17-19, 28-30
dipwells 81-82, 124-136, 191-192, 197, 199
ditches
　blocking 162-163, 199, 201, 213
　drainage, 68, 160-163, 173, 176, 198-199, 201, 213
Djibouti 51

drainage 2-5, 15- 17, 21, 24-27, 46-48, 51, 54-57, 62-71, 75-84, 90, 92, 103-119, 128, 130-136, 139, 141, 143-, 155-168, 170-171, 173-194, 198, 199-215, 221

ecohydrology 1, 9-12, 102
Environmental Impact Assessment 29
erosion, soil 4, 22, 26, 52-53, 111, 132, 146-147, 157, 170, 194, 213, 220-221
Ethiopia 4-7, 14, 18, 20, 50-61, 65, 188, 205, 218, 220
evapotranspiration 10-11, 16-17, 20, 24, 26, 102
EWRP 54, 95, 100-111, 149
experimentation 23, 35, 36-40, 48, 92-93, 172, 176, 179
extension agents 35, 38

feudal system 60
fishing 18
flooding 13, 16, 18, 21-23, 79, 102, 105, 108, 117, 130, 146, 152, 162, 170, 206, 213, 215
floodplains 17-19, 22-23, 27

geology 10, 16, 56
GIS, 77, 79, 98
globalization 3
Gore 56, 62, 84, 122-124
government policy 67, 221
grazing 5, 18, 23-26, 47, 62, 65, 67, 69, 77, 92, 103-109, 112-117, 126, 130, 133-134, 144-148, 155, 159, 165, 210
green revolution 32-33, 47
ground truthing, 79, 98-99

Hadejia-Nguru wetlands 23, 27
Haile Selassie 59-60, 106, 111-114, 117, 119, 219
hydraulic conductivity 16, 80, 83, 130, 134, 145, 147, 191, 193, 195

hydrological monitoring 75, 80-83, 94, 100, 119, 147, 179, 182, 188, 192-196, 203

Illubabor 4-7, 51-76, 94-95, 100, 106, 108-109, 115, 122, 149, 167, 172, 181, 205, 218-222
indigenous knowledge (IK) 4-7, 30, 34-50, 73, 80, 94, 124, 149, 172-173, 176-181, 194-195, 206, 216-222
innovation 36- 39, 47, 172-173, 176, 179, 216, 220, 221-222
irrigation 3, 18, 23-26, 30, 39, 46, 54, 57, 66, 68

Jimma 54, 218

Kafue Flats 23
Kenya 16, 22, 51, 220

landlords 117
loop concept 42, 218

maize 17, 25, 57, 60, 63-68, 77, 82, 104-109, 112, 115, 117, 133, 152, 158, 159-160, 163-164, 167-168, 175, 192-193, 198-201, 213
Malawi 48
market economy 66, 178, 219
medicinal plants 18, 28, 40, 64, 71, 92-93, 106, 112, 159, 168, 220, 222
Menelik 59, 109
Menschen für Menschen (MFM) 63, 68-71, 75-76, 109, 111, 175, 207
Metu 58-63, 66, 75, 84, 100, 105-106, 108-111, 114-115, 123-125, 219
migration 4, 59, 169
mineralization of soil 132, 145, 194, 213
Ministry of Agriculture 54, 59, 61-69, 75-76, 93, 174-175, 177
Ministry of Tea and Coffee Development 65
Ministry of Water, Minerals and Energy 61, 66, 75-76
multiple use, of wetlands 115, 159, 210

NGOs 48, 63, 68, 175
Niger 22-23, 38
Nigeria 18, 23-27, 41
Nile, River 20, 53, 54, 61
nitrate 12, 83, 137-144

Oromo peoples 59, 168
oxidation, of soil 22, 82-83, 145, 147

peatlands 11, 14
pH 56, 82-83, 114-117, 137-139, 146
phosphate 12, 83, 137, 141-144
ploughing, in wetlands 104, 133-134, 143, 160, 212-215
population pressure 3-4, 21-23, 53, 59, 65, 71, 220
PRA 85-95, 100, 149, 150-165, 170, 172-175, 181-182, 185, 188, 192, 199, 201-202

rainfall 14, 18, 20-26, 53, 56, 58, 61-64, 67, 80, 84, 89, 91, 93, 102, 106, 122-132, 137, 139, 144-146, 150-152, 156-157, 176, 178, 182-185, 188, 194-197, 201-203, 213, 220
Ramsar Bureau 27-28, 49
regeneration, of wetlands 28, 104, 119, 146-147, 169-170, 192-195, 203-210, 214, 219
resettlement 4, 59, 65
rice 17-18, 21, 23-25, 37, 47, 48
Rwanda 20, 62

scarecrows 177
scientific knowledge 35, 41-45, 49, 89, 195, 222
seasonal diagrams 89, 93, 150, 152, 155-157, 182, 188
semi-structured interviews 93
Sierra Leone 21, 24, 37, 47
Somalia 51
sowing 104, 160, 163, 199
study area 6, 51, 75-79, 80, 84, 94, 95-101, 106, 118, 122-125, 137, 152, 179, 181, 188, 204, 207, 210, 216, 219, 221-222
Sudan 18, 20, 22, 51, 53-54, 64
Sudd, region of 20, 54
swamps 1, 4, 13, 19, 20-26, 37, 46-47, 54, 62, 102, 132

Tanzania 20, 21, 25
Tools 68, 90-93, 164, 178
transfer of technology approach 32, 38, 46
tukul 165

Uganda 20-22, 29, 46, 62, 71, 132

venn diagrams 92-93, 173-175
villagization 66, 108, 169

water table 11, 13-16, 20-26, 62-71, 80-84, 89, 92, 94, 102-104, 124-136, 144-157, 165-166, 170, 178, 182, 185-206, 210-214

weeding, practice of 104, 160, 164, 214
Wetlands International 49
wild pests 25, 65, 117-118, 160, 164, 169-171, 175, 176
wildlife 17-20, 22, 27-29, 53, 71

Zambia 17, 19, 23, 46
Zimbabwe 18-19, 25